Gandhi's Printing Press

GANDHI'S PRINTING PRESS

Experiments in Slow Reading

ISABEL HOFMEYR

HARVARD UNIVERSITY PRESS

Cambridge, Massachusetts, and London, England

2013

Library of Congress Cataloging-in-Publication Data
Hofmeyr, Isabel.
Gandhi's printing press : experiments in slow reading / Isabel Hofmeyr.
p. cm.
Includes bibliographical references and index.
ISBN 978-0-674-07279-4 (alk. paper)
1. Gandhi, Mahatma, 1869–1948—Political and social views.
2. Indian opinion (Durban, South Africa)   3. Reading—Political aspects.
4. Newspaper presses—South Africa—History.   5. Newspaper publishing—South
Africa—History.   6. Printing industry—Indian Ocean Region—History.
7. Great Britain—Colonies—Public opinion.   8. East Indians—Attitudes.   I. Title.
DS481.G3H53 2013
954.035092—dc23   2012031196

In Memory of
Miriam Abrams
(1958–2011)

Contents

GANDHI'S PRINTING PRESS

Introduction

A Gandhian Theory of Text

In one of his many memorable phrases, Benedict Anderson describes imperialism as a process of "stretching the tight skin of nation over the gigantic body of empire."[1] To Mohandas Gandhi, a reluctant nationalist at best, this sentiment would have seemed back to front. For much of his life, Gandhi regarded empire as cardinal, the nation-state as secondary, an unfortunate growth upon it. Yet as nationalism gained ground, the equation had to be presented the other way around. What happened when one tried to bunch the vast skin of empire on the nation? What to do with all those ungainly folds?

Gandhi's political project, especially during his South African years (1893–1914), can be interpreted as an attempt to smooth out some of those gathering folds. Whether espousing imperial citizenship or what from the 1920s would be called Greater India, Gandhi grappled with how to articulate versions of "India" and "empire" (understood as a guarantor of rights for British Indians) that could be interchangeable.[2]

In a radical formulation, Gandhi devised one answer to this problem by inserting "Truth" (or satyagraha/passive resistance/soul force) as the third term in the equation. The resulting triad—Truth, India, Empire—compelled his imagination; to it, he "own[ed] allegiance," as the preface to the English version of Gandhi's manifesto *Hind Swaraj* explained.³ By being a true satyagrahi, anyone, wherever they were, could learn to rule the self, creating a miniature zone of sovereignty and thereby laying one small precondition for broader ideals like India and empire.

This equation located self-rule primarily in the individual rather than in a territory. This novel idea of sovereignty took shape in the context of southern Africa where, on the one hand, Gandhi had to endure the obscenities of white settler nationalism and, on the other, ponder who could be Indian outside India. The elegantly minimal definition of India and empire that these circumstances produced did not survive the journey back to India. In a climate of anti-imperialism, nations are destined to become more national, erasing the traces of their prefatory formation elsewhere.

This book concerns itself with these utopian traces of Gandhi's South African years. Gandhi's printing experiments during his South African sojourn provide the lens for doing so. These comprised a printing press, the multilingual newspaper *Indian Opinion,* and the pamphlets that it produced (the most famous of which is *Hind Swaraj*). Opening its doors in Durban in 1898, the International Printing Press (IPP) (of which Gandhi was a sometime proprietor) consisted of a small jobbing operation that from 1903 began printing *Indian Opinion*. A year and a half after the launch of the paper, the IPP moved to Phoenix, Gandhi's first ashram, fourteen miles north of the city. From here the IPP continued to produce *Indian Opinion,* and both the press and the paper found themselves protagonists in the larger story of satya-graha, or passive resistance, in South Africa, which unfolded between 1906 and 1914.

This Gandhian textual culture is worth investigating not only for its own sake but also for the light it throws on the philosophy of satyagraha, popularly understood as a political program of nonviolent resistance or noncooperation. Yet satyagraha can equally be understood as a practice of sovereignty rooted in the individual rather than as an abstraction like "nation" or "movement." Seen in this way, satyagraha inheres in forms of conduct (like spinning and fasting) through which ethical values are accreted in the satyagrahi, as Uday Mehta has brilliantly shown.[4] Such practices form the basis of self-rule or ruling the self, creating sovereignty, one person at a time. Reading and writing can be made to conform to such practices of the self and can hence be a mode of satyagraha as well as an activity through which to theorize the idea itself.

In the Indian Ocean port city of Durban, Gandhi and his colleagues experimented with the ways in which printing and publishing could enlarge new kinds of ethical selves. Situated on an ashram (or an ashram-like settlement), the printing press on a daily basis enacted a novel order of community, drawing in different castes, religions, languages, races, and genders. Gandhi progressively phased out advertisements from the newspaper and took little notice of copyright legislation, constructing an ideal reader freed from the addictions of the markets and the dictates of the state. As a paper addressed to audiences in Britain, Africa, and India, *Indian Opinion* explored ideas of "India" that were not territorially based but, rather, existed among the individual sovereignties of its readers and the pathways of circulation that linked them.

These experiments unfolded in an age of vertiginous acceleration via trains, steamships, and telegraphs, where, with mounting intensity, an industrialized information order bombarded readers with more and more printed matter. Ever-briefer media genres like the headline, summary, and extract speeded up tempos of reading. In Gandhi's view, such reading "macadamized" the mind (to use an image from Thoreau that appeared in *Indian Opinion* on June 10, 1911) and reinforced the dangerous equation of speed with efficiency.[5]

Gandhi, by contrast, sought to slow down reading and textual production more generally. He favored hand printing and encouraged a style of reading that was patient, that paused rather than rushed ahead. He interspersed news reports with philosophical extracts, and he encouraged readers to contemplate what they read rather than to hurtle forward. In effect, he experimented with an anti-commodity, copyright-free, slow-motion newspaper.

In exploring these ideas, Gandhi worked with two obvious and everyday truths, namely, that serious reading can only be done at the pace of the human body and that each reader must read on his or her own behalf. If we are to read thoughtfully, we cannot speed up the pace at which we read and we cannot outsource the activity to someone else. In a Gandhian world such slow reading became one way of pausing industrial speed, and in so doing it created small moments of intellectual independence. Reading might happen within the world of industrialized time but did not need to be entirely driven by its logics. This focus on bodily rhythm as a way of interrupting industrial tempos became central to Gandhi's larger and world-famous critiques of modernity that questioned the equation of speed with efficiency and technology with progress.

Ways of Understanding Gandhian Text-Making

In telling the story of Gandhi's experiments with reading and writing, this book seeks to encourage a more historicized approach to the topic. Most discussions of Gandhi's textual production start from the assumption that the genres and forms he used more or less resemble our modern-day equivalents. So when discussing Gandhi's publishing activities, most scholars assume that *Indian Opinion* is a "newspaper" and Gandhi an "editor" (if only de facto—he never technically edited the paper, although he was always its senior figure). Together they produced standard journalistic genres like the editorial or the news report. Yet once one looks closer, this familiarity evaporates. Much of Gandhi's contribution to *Indian Opinion* took the shape of ethical extracts from "great men" like Thoreau, Tolstoy, and Ruskin. What initially looked like a newspaper starts to resemble an ethical anthology. We search in vain for stories from reporters on the beat or accounts filed by foreign correspondents. Instead we see the standard fare of the Victorian and Edwardian periodical, namely, the clipping, summary, or extract from another publication.

Likewise, in the world of printing, the categories we might expect—compositor, machine man, foreman, manager—are blurred, with staff at the IPP regularly crossing these frontiers. In addition, this text-making world operated multilingually. In its early years, the IPP offered printing services in ten languages (English, Gujarati, Tamil, Hindi, Urdu, Hebrew, Marathi, Sanskrit, Zulu, and Dutch), involving seven different scripts (*Indian Opinion*, June 4, 1903). *Indian Opinion* initially appeared in four languages (English, Gujarati, Tamil, and Hindi) and four scripts, probably an entirely novel experiment since most vernacular newspapers in India itself used, at most, two languages. Under such circumstances, there were no fixed or predetermined ways of working. Instead, processes of production had to be invented to accommodate the polyglot operation and the work of translation that formed part of it.

There are further examples of such retrospectively applied notions. These include the idea of Gandhi as an "author" who produced "books." On this point, one interesting piece of Gandhiana would be to establish when his work first appeared between hard covers. Since many early editions of his work do not survive, the answer will always be speculative. If pushed, I would put my money on the first edition of *Satyagraha in South Africa* (produced in 1928 by Ganesh Publishers in Madras) or possibly an edition of *Hind Swaraj* from the early 1920s. Gandhi's first piece of public writing appeared as a series in *The Vegetarian* in 1889 (*Collected Works* [CW] 1: 18–40).[6] It hence took some three to four decades for his work to migrate from the lower reaches of serial print to the higher echelons of the "monumental" book.

The template of individual "author" and "bookness" can be misleading, as evidenced by the otherwise admirable *A Comprehensive, Annotated Bibliography on Mahatma Gandhi* that lists 497 entries under the heading "Books by Gandhi."[7] Yet rather than being books by one author, they are editions, compilations, abridgments, collections, anthologies, selections, arrangements, translations, pamphlets, multivolume works, speeches, interviews, prayers, hymns, and songs. Fortunately the word "by" brings its full prepositional and adverbial range to the rescue. These "books" are literally "by" "Gandhi": they are near or next to him; they travel "by" him as one travels by air or by bus; they happen in the temporality of his life and afterlife (by day, by night, by "Gandhi"); they emerge by Gandhi's account, evidence, and authority; they function according to and in conformity with "Gandhi." The Gandhian book indeed emerges "by and by."

Most analyses of the Gandhian book take India as their primary focus and the textual practices Gandhi formulated there as the norm. After his arrival in India, and as the scale of his publication projects expanded, Gandhi's views on copyright shifted, especially as the reality of the large-scale international publication of his autobiography forced him into a limited and pragmatic use of copyright.[8] This reluctant acceptance of copyright was accompanied by a growing formalization of his work via the Navijivan Trust, established in 1929, to which Gandhi ceded his copyright in 1940. In 2008 the copyright on Gandhi's work expired and the Trust, despite pressure to do otherwise, refused to extend the copyright, citing Gandhi's ambivalence toward intellectual property legislation.

Much Gandhian writing is understood looking back from this vantage point where his work appears in copyrighted book form. The experimental character of his early textual production in South Africa is hence obscured. This study suggests that we look more closely at the textual conditions under which Gandhi worked in South Africa.

Gandhi's textual experiments unfolded initially in Durban, an obscure port city, at first glance an unlikely setting for the emergence of his world-famous ideas. Yet like equivalent entrepôts around the Indian Ocean littoral, Durban was home to a jumble of diasporic communities from different parts of the world, an environment that obliged its inhabitants to experiment with who they were or could become. Part of the vast imperial migrations of the nineteenth century, these flows of indentured laborers, migrant workers, prisoners, sailors, pilgrims, merchants, and missionaries coalesced around the Indian Ocean littoral, in port cities (and their hinterland capitals) like Cape Town, Johannesburg, Zanzibar, Mombasa, Nairobi, Port Louis, or Bombay.[9] In these destinations, Africans, Middle Easterners, Southeast Asians, Chinese, and Indians found themselves as happenstance neighbors, relating both to a colonial authority and to each other.

These zones of enforced cosmopolitanism were not simple melting pots but, rather, arenas that crystallized new lines of allegiance, belonging, and exclusion. People had to distinguish themselves from their colonial rulers, from their new "also-colonized" neighbors (to use an apt phrase from Jon Soske), *and* from those who were nominally the same.[10] Muslims from various regions of India encountered a colonial state, African neighbors, and other co-believers from the subcontinent. The languages, dress, food, and theologies of these coreligionists were at times so different as to make them almost as strange to each other as they were to the new communities among which they found themselves.[11]

The epic mobility of nineteenth-century imperialism, of which these kaleidoscopic communities formed a part, engendered a rich array of transnational imaginings, as displaced and dispersed communities had to envisage their position in a new order. With its deep history of maritime mobility, the Indian Ocean world fostered an especially rich array of transnational associational forms, as scholars like Sugata Bose, Mark Ravinder Frost, T. N. Harper, and Engseng Ho have demonstrated.[12] In these port cities, Arya Samajists, theosophists, Sufis, pan-Buddhists, Sikh transnationalists, African nationalists, white laborists, and those espousing imperial citizenship established transoceanic organizations speaking in global idioms.[13] Exploiting the improved communications networks of empire, these groups created and sustained far-flung memberships. They sent out evangelists, circulated pamphlets, and undertook lecture tours, often sharing expertise and technology with like-minded organizations. Key elements in the organizational grammar of these movements were port cities and print culture.[14] Through steamer, telegraph, and railway, these entrepôts functioned as relay stations for the texts, commodities, and personnel of the dispersed groups.

While these transoceanic projects evinced cosmopolitan elements, they equally constructed distinct boundaries whether of "race," civilization, religion, or class. The universal dimensions of such projects aimed both to appeal to a wide audience and to trump the white supremacist claims of empire. Arya Samajists and Indian nationalists presented the ancient diffusion of Hinduism as an instance of Greater India, an example of a benign "empire" that showed up the shortcomings of Greater Britain.[15] Yet at the same time such claims ventriloquized a set of more parochial and sectional interests. Rochelle Pinto's analysis of print and politics in Goa brilliantly demonstrates how the nineteenth-century Goan elite used print technology to feint at Christian universalism while insinuating caste into print and the historiography of Catholicism.[16] *The Hindi,* a short-lived bilingual paper of the early 1920s in Natal, aired the gossip of the tiny Hindi- and English-speaking elite alongside extravagant claims of the paper's apparently global reach. The newspaper's masthead advertised the publication as "A HIGH-CLASS AND FEARLESS JOURNAL OF POLITICS, LITERATURE, COMMERCE, INDUSTRY, RELIGION & SOCIAL REFORM . . . Circulated throughout South & East Africa, British & Dutch Guiana, Mauritius, Fiji, America, Europe & India."

Gandhi likewise espoused ideas that were universal but bounded. His ideas about India were especially capacious. Their core focused on the diverse collection of Indian communities thrown together in South Africa. This assemblage of peoples made a miniature version of India visible, with a clarity not conceivable in the vastness of the subcontinent itself.[17] Beyond this core, Gandhi's ideas of India stretched to include even those who spoke no Indian languages or were not of Indian descent. These groups could belong, provided they apprenticed themselves to Indian civilization, an idea that is explored in Chapter 4. Africans, however, were not included (or at least so Gandhi thought during his South African years—his views were later to change).[18] Influenced by hierarchical ideals of civilizationism, Gandhi defined Africa as outside the pale of India and empire. In so doing, he installed Africa as a boundary of India.

Such use of the "also-colonized other" as a frontier of definition was commonplace. Proponents of African nationalism likewise took "India" as a boundary in terms of religion (Christian Africans versus "heathen" Indians), indigeneity (sons of the soil versus Indian settlers), and commerce ("intruding" Indian traders taking over "African" markets).[19] The arm's-length relationship between Gandhi and his neighbor, John Dube, the first president of the African National Congress and head of the Ohlange Institute, located next door to Phoenix, embodies these frontiers. Each involved in creating his own miniature "continent," the two men defined themselves in opposition to each other, admiring each other's projects from afar but deprecating each other's "people"— Gandhi being as well known for his anti-African statements as Dube for his anti-Indianism (even while his newspaper carried advertisements for Indian merchants).[20]

It is of course easy to condemn such projects with the censoriousness of hindsight. Yet, as Tony Ballantyne has argued, we need to place them in a broader context of imperial race-making in which ideas about race do not emerge solely from Europe but are constructed by a range of intellectual players and groups across empire.[21] Such race-making projects were common in an age where ideas of ethnic regeneration and Booker T. Washington-style segregational self-sufficiency were habitual. While anticolonial alliances across colonized groups certainly existed, these sentiments were less common than a later age of decolonizing anti-imperial solidarity, third world nationalism, nonalignment, and Afro-Asianism might make us believe.

This book seeks to locate Gandhi and his printing experiments squarely within this contradictory and confusing world. The term I use to capture the complexities of these multiple intersections is "colonial born," a now virtually obsolete word that described Indians born in South Africa. This term formed part of a rich lexicon for designating Indian communities outside India—"Indians overseas," "Indians abroad," "Indian colonials," "Greater India," "Indians in the dominions," "Indians outside India," and the "Indian emigrant" (the last also the name of a Madras-based newspaper aimed at the diasporic world). This vocabulary has, however, been eclipsed by the blanket term "diaspora," a post-1960s word popularized in discussions of the more middle-class migrations to Europe and the United States that changing immigration policy occasioned.[22]

The ambiguity of the term "colonial born" is useful for our purposes. In general, the phrase denoted those born in indentured destinations like the Caribbean, Fiji, or Natal, and technically the term is correct—these were colonies. Yet the expression equally implies that India itself is not a colony, and that it occupies the position of metropole or mainland vis-à-vis its indentured and somehow "colonial" periphery. This book adopts the doubleness of the term "colonial born" as an analytical concept that can do useful work in simultaneously illuminating the entanglements of the vertical and horizontal linkages of empire. This double meaning of the term allows us to apply revisionist postcolonial debates on empire that argue for treating "center" and "periphery" as one integrated space *both* to a metropole/colony axis *and* to India's subimperial relations with its indentured peripheries.[23] The term "colonial born" hence captures the idea of an entanglement of colonial, semicolonial, para-colonial, and anticolonial formations that took shape at the point where Greater Britain and Greater India intersected in southern Africa.

As Chapters 1 and 2 describe in more detail, the printing press that Gandhi established took shape in these entanglements: it drew its personnel from southern Africa, the Indian Ocean world, and England, and it sought to address an audience spread across Africa, India, and Britain. I hence describe the press and paper that Gandhi set up as being "colonial born" and explore the ways in which Gandhi exploited the opportunities inherent in its contradictions. If the medley of diasporic communities resulted in varied printing personnel, then this disparateness provided the basis for trying out new forms of community. If dispersed audiences were to be addressed, then this exigency prompted experiments in creating a deterritorialized India as well as new kinds of horizontal textual alignments between South Africa and India.

In pursuing these experiments, Gandhi used the standard media forms of the age, namely, the pamphlet, the periodical, and its primary building blocks—the cutting, summary, and abridgment taken from other publications in keeping with the widespread exchange system by which consenting journals agreed to cannibalize each other's material. A closer understanding of the forms and genres that he employed is essential, since it is from them that Gandhi, ever an alchemist of the ordinary, generated some of his key ideas.

Periodical, Pamphlet, and the Limits of Print Capitalism

In relying on the periodical and the pamphlet, Gandhi was not unusual—these were the forms of empire par excellence. A cut-and-paste assemblage of publications from elsewhere, the periodical on every page convened a miniature empire. Pamphlets, which often arose from the pages of periodicals, were a major medium for making ideas more portable and durable across the empire. In part, financial constraint accounts for this prevalence of the periodical and pamphlet: daily newspapers and hardback books presuppose considerable resources in the way that weekly periodicals and leaflets do not. Yet beyond these pragmatic concerns lie broader themes that direct our attention away from the forms that normally occupy such discussions—the newspaper and the novel—and toward the more humble periodical and the pamphlet.

As Anderson has taught us, the simultaneous consumption of newspapers and novels by consumers encourages a sense of shared nationhood. But daily newspapers are designed to circulate in circumscribed areas—generally a city or, at times, the nation. Periodicals, by contrast, are not tied to one place and are intended to circulate widely.[24] Their weekly, monthly, or quarterly tempos are slower than the daily rhythms of the newspaper but, more importantly, are driven by the temporalities of circulation ("By the time this periodical reaches Calcutta . . ."). Reading such periodicals can never be a simultaneous experience as with the daily newspaper; rather, their modes of consumption are punctuated and sequential. A further difference is that newspapers are rapidly consumed and cast aside. Periodicals have longer shelf lives as they meander to their different audiences. Their modes of production are less date-driven, and they rely more on the undated excerpt, essay, or clipping, taken from other periodicals.

This distinction between a periodical and a newspaper may sound a bit overdrawn: one came out daily, the other weekly, monthly, or quarterly, but in appearance they resembled each other and borrowed materials from each other. Both belong to a bigger category of serial print culture whose scholars have long given up the task of trying to draw neat distinctions between different categories of periodical publication.

Yet the distinction in this instance is worth making, not least because the newspaper is so central in Anderson's analysis. Here a newspaper is a daily occurrence, discarded after reading, bought as a commodity in a marketplace, and read privately in one language.[25] Saturated with "commodity-ness," the newspaper represents the end point of a world where the print industry emerged as one of the earliest forms of capitalist enterprise, inducting textual objects as one of the first full commodity forms.

By these measures, *Indian Opinion* was hardly a newspaper at all. It appeared weekly, was read both privately and communally, was stored up for future use, appeared in at least two languages, and published news as well as more durable essays and ethical discourses. Such periodicals were never solely objects for profit but formed part of a world of philanthropy, reform, and service. Pamphlets too were rarely straightforward secular commodities: seldom intended to render profit, they often related religious plots of eternal life rather than secular narratives of national time like the novel.

Christopher Reed's discussion of Chinese print capitalism is apposite to this broader debate. He sees print capitalism as having arrived when "commercialized, secularized, nongovernmental, and nonphilanthropic printing came to be done, not as a handicraft, but as a form of industry carried on by machine."[26] These circumstances hardly apply to small "colonial-born" operations like Gandhi's printing press: barely commercial, semisecular, generally nongovernmental, and always philanthropic, these presses were artisanal operations, not industrial plants, in Reed's sense.

With themes of philanthropy, merchant patronage, personalized address in a transnational matrix, artisanal production, multilingualism, and variable notions of authorship and copyright (a feature taken for granted in the world of print capitalism), these presses invite us to relativize the term "print capitalism" with its themes of intellectual property regimes, machine-driven nationalism, and vast, monoglot publics.

Some alternative terms may be appropriate. "Vernacular print capitalism" usefully problematizes the phrase yet captures the dimensions of multilingualism and popular idiom. Sheldon Pollock's use of the phrase "script-mercantilism" to describe South Asian systems of manuscript production suggests the expression "print mercantilism," and while not intending to imply that these presses were run by merchant elites, the phrase does capture the element of philanthropic commerce, family ownership, and the diasporic routes that shaped them.[27] Karin Barber's term "printing culture" captures the small artisanal qualities of these presses better than the term "print culture," with its implication of vast, anonymous, and secular address.[28] One could also make recourse to terms like "print Islam," "print Samajism," "print laborism," and so on, all of which remind us that it was these ideologies and their intersections, rather than the technology itself, that shaped the meaning of printing in diasporic settings.

The Building Blocks of the Periodical

At least in pre-telegraph days, the imperial periodical press ran on an exchange system in which consenting journals swapped publications from which they could garner material. This scissors-and-paste journalism consisted of cuttings, quotations, extracts, précis, and abridgments of material from elsewhere.[29] In some parts of empire, this system persisted into the post-telegraph era, especially in contexts where publications could not afford expensive wire service fees, objected to their imperial bias, or stood outside the imperial press system that united Britain and its "white" dominions via flows of personnel, funding, technology, and supplies of international news.[30] Periodicals fashioned from these exchange papers carpeted empire, creating endless textual intersections that constituted part of the fabric of imperial life itself.

Summary, Speed, Subjectivity

One important periodical staple was the summary, and from this lowly genre Gandhi elaborated radical ideas. Summary saturated the world in which he lived. First, much of his day-to-day routine entailed summary—abridging stories from other papers; summarizing laws, statutes, and policy documents; undertaking condensed translations of key works; and drafting precise memorandums and petitions. For him, summary constituted a seminal skill, the "art of condensation" (*CW* 10: 195), as he called it, and he explored all of its modes and shades—précis, abridgment, adaptation, and epitome, as well as the pamphlet, another miniature form. Indeed, his characteristic prose style—always pared to the bone—arose out of this labor of sifting and sorting, refining and honing. Replete with detachable maxims, Gandhian prose has provided rich fields for harvesting the sayings of the Mahatma: *Gandhigrams, Gandhi's Wisdom Box, A Gandhian Rosary, Precious Pearls: Glittering Galaxy of Gandhian Gems, Thus Spake Mahatma,* and so on.[31]

The second way in which summary saturated Gandhi's world was through the information flows of empire where intelligence from afar generally arrived as a summary of a summary, a condensation of a condensation. Imperial modernity relied on summary, telegram, telegraph, clipping, and extract for its operations. Indeed, Gandhi was to experience firsthand the consequences of such "telegraphic imperialism."[32] On a trip to India in 1896 a three-line condensation of his Green Pamphlet (that depicted the disabilities faced by Indians in Natal) was telegraphed to London and then on to Natal, where white settlers used the précis to aggravate a populist climate that would result in Gandhi nearly being lynched on his return (Fig. 5).[33]

As Gandhi's understanding of colonial and imperial rule widened to take in industrial capitalism as an explanatory frame, he equally came to appreciate these flows of summary as tied to the industrialization of information itself. As ever faster and more epitomized bits of information bombarded consumers, reading itself became calibrated to machine-driven rhythms.

The resulting Malthusian textual order precipitated widespread international debate on changing reading patterns. In Victorian thinking, reading was closely tied to ideas of character (or subjectivity, as we would now term it); changes to how people read could impact on what kind of subjects they would become.[34] Commentators expressed dismay at a situation where dramatically increased volumes of print turned reading into an indiscriminate, addictive, incoherent activity in which people became machine-like. Headlines, billboards, street advertisements, and short news articles instituted rapid ways of reading that accommodated themselves to the urgent rhythms of metropolitan life. For many observers, such superficial reading could only be debased and was judged against supposed norms of sedate and sustained reading.[35]

Gandhi's textual experiments recognized and accepted this dispensation of discontinuous reading as an ineluctable condition of modernity and its information orders. His response was not to sneer at it or to try and upend this system but, rather, to recalibrate it from within by fostering practices of "slow," syncopated reading.

In the world of Victorian and Edwardian periodicals, these ideas were not unusual. In an environment preoccupied with accelerating production and rhythms of reading, Gandhi's ideas would have been familiar, especially among the more socially conscious "old journalism" Edwardian periodicals that *Indian Opinion* resembled and to some of whose representatives it had links (*New Age* being one example [*Indian Opinion*, April 2, 1910]).[36]

These various tempos of "industrial reading" are now recognized as powering the literary forms of modernity itself and have been explored from different angles: the rhythms of serialization and the novel as a genre; little magazines as syncopated hothouses for modernity; and anthologizing as a method of training in selective reading.[37] While Gandhi's experiments with reading and time can usefully be bracketed into this field, they equally outstrip it by placing reading in a wider transnational frame and by being more alert to the centrality of circulation.

These models of reading that Gandhi forged took shape in the context of the periodical. Made up of clippings from elsewhere, each periodical is marked by the residue of prior readers as well as the hypothetical knowledge of those to come. Reading is hence an experience of witnessing a provisional thickening of text pulled from across the world cohering briefly, before dispersing again into a range of other periodicals.

If periodical reading constituted an interaction with a conveyor belt of text, then the reader necessarily had to function as an editor/redactor. Such practices of reading/editing became one index of how to gain sovereignty within the potential pressures and tyrannies of circulation. Through patient reading, through a careful selection of texts, through mentally inserting ethical extracts into hasty news items, and through resisting macadamization, readers could slow down the system, turning themselves into nodes of autonomy not through abstract ideals but through these small daily textual practices. Such practices of reading/editing constitute a realm of public opinion as sovereign not via an abstract assertion of commonality based on the simultaneous consumption of texts à la Anderson but, rather, through this slow craft that proceeds by punctuated sequence, both within individual readers and among different readers.

These ideas of reading and sovereignty intersect most visibly in *Hind Swaraj*. Although produced during his South African years, Gandhi's manifesto is routinely treated as a text about the future of India addressed to various anticolonial revolutionaries in England and the subcontinent whose violent methods he abjured.[38] Yet the early prefaces of the book explicitly indicate that the pamphlet is aimed at readers of *Indian Opinion,* something that has almost universally been ignored. The analysis presented here reinstates the reader of *Indian Opinion* not to supplant the existing analyses or imply they are wrong but to suggest that the two imagined audiences (the reader of *Indian Opinion* in South Africa and the revolutionaries in England and India) need to be considered together.

The book offers us a dialogue between an Editor and a Reader during the course of which the Editor trains the Reader to read and interpret correctly. This process shifts the Reader from his misguided admiration for the violent methods of the revolutionaries toward becoming a potential satyagrahi. To see this dialogue as relating exclusively to India (or only to the South African world of *Indian Opinion* for that matter) is to miss the point. Rather, both interpretations are correct, given that "India" exists as a deterritorialized space created through textual circulation in which virtually anyone, anywhere, could make themselves Indian. The form of *Hind Swaraj* that unfolds in the imaginative context of a newspaper but has no specified territorial setting underlines this point.

Since this Gandhian theory deals with reading, it is necessarily elite. Indeed, the ideas that Gandhi formulated about

"India" in South Africa were largely aimed at those who could read. As the numerous accounts of satyagraha have indicated, in the very final stages of his campaign he did draw in illiterate indentured workers. However, these workers acted on their own agenda with regard to working conditions rather than any idea of India's national honor, a concept whose currency was restricted largely to an educated elite.[39]

Yet readers are both real and imagined. The focus on reading, writing, and textual production as daily and continuous crafts was both literal and metaphorical and, in this latter guise, resonates with Gandhi's later thinking about noninstrumental activity. While it is of course difficult to imagine reading or writing that is noninstrumental, Gandhi's thinking at its very limits asks us to ponder what such an activity might look like.

In raising this question, this book engages with political theorists whose work offers us new vistas on Gandhian thought. As scholars like Uday Mehta, Ajay Skaria, and Tridip Suhrud have argued, Gandhi is a thinker rooted in the singular and everyday rather than in abstract principles, instrumental logic, and theoretical teleologies.[40] As Mehta points out, Gandhi focused "on the materiality and minutiae of everyday life, where what matters is the tactile immediacy of the moment." His favored activities of spinning, fasting, and celibacy represent "self-conscious withdrawal" from the pressured trajectory of means and ends.[41]

Skaria has extended these arguments to Gandhi's ashrams, "the conceptual site through which he tried to think about and develop practices for an alternative politics."[42] Central to the vision of the ashram was a concrete focus on the neighbor and one's obligations to him or her. As Skaria argues, such practices moved politics away from the abstract seductions of the nation-state and toward the singularity of specific interactions.

Gandhi's Printing Press extends these ideas through telling a story of a printing press on an ashram very far away from India. In such conditions, Gandhi necessarily had to confront an abstract and vicarious politics: "India" was a virtual entity some four thousand miles distant; ideas of "whiteness" claimed authority through abstraction (law, rights, civilization). The idea of the neighbor or neighborliness was likewise not fully formed: Gandhi had in fact abandoned his neighbors in Gujarat, conforming to a type that he was subsequently to castigate: "a man who allows himself to be lured by 'the distant scene' and runs to the end of the earth is not only foiled in his ambition but also fails in his duty towards his neighbours . . . [this] decision would throw [the] little world of neighbours and dependents out of joint, while [this] gratuitous knight errantry would . . . disturb the atmosphere in the new place."[43] Likewise, Gandhi had neighbors like John Dube with whom he wanted little to do.

These complications indicate that we need to think about Gandhi not only from the point of view of the subcontinent but in the triangle of India, South Africa, and England. Such a configuration reminds us again that Gandhi worked at the intersection of imperial and subimperial and colonial and anticolonial axes. Holding on to all these dimensions makes visible the forms of his South African experiments that he continued in India while also bringing into focus those strands of Gandhian thinking that have been lost. It gives us a renewed appreciation of the spectrum of Gandhian thought and hence of his genius and innovation.

Outline of the Book

This book tells the story of a press, a paper, some pamphlets, and the idea of the reader that these engendered. Chapter 1 describes the Indian Ocean arenas of print culture from which the IPP and *Indian Opinion* emerged. This discussion focuses on the phenomenon of small printing presses around the Indian Ocean world that took shape in multiply diasporic contexts. Chapter 2 narrates a biography of the IPP in two parts: the first section focuses on the early days of the press in Durban, paying attention to its diverse personnel, and how they set about running a multilingual, multiscript, "colonial-born" operation. The second section tells the story of a press and an ashram, Phoenix, whence the press moved in 1904. Here the press took on even stronger utopian attributes, which were in part unique but also mirrored similar printing projects by Africans, Muslims, and Hindus in the Phoenix hinterland.

Chapter 3 shifts to *Indian Opinion* itself, paying attention to the newspaper as shaped in the vast "exchange paper" circuit of which it formed one node. I examine the ways in which staff used the idea of circulation to critique colonial ideas of time and space while also opening up new intellectual pathways between South Africa and India. Gandhi extended this exchange system into the realm of ethical writing, weaving in extracts and summaries of writers like Thoreau, Ruskin, and Tolstoy among news stories. In juxtaposing these forms, Gandhi asked readers to equate them, turning the larger temporalities of ethical writing against the date-driven rhythms of news while also asking what ethical news reporting would be. Such strategies constituted part of a larger climate in the paper that sought to deinstrumentalize time and hence slow down reading.

Chapter 4 discusses the thirty-odd pamphlets that the paper generated. These constitute an extraordinary intellectual lineup that includes Socrates; the Egyptian nationalist Mustafa Kamal Pasha; Ruskin; Annie Besant; J. L. P. Erasmus, a Boer commandant (who provides an English retelling of the Hindu epics); Tolstoy; and the politician, historian, and jurist Syed Ameer Ali. The chapter binds these pamphlets together to examine the way in which they explore different ideas of India outside of India itself, a series of hypothetical experiments in Gandhi's attempt to establish a "summary India" in South Africa.

Chapter 5 turns to an examination of the ideal reader that Gandhi constructed, mainly in the Gujarati columns. (Since this figure is implicitly imagined as male, I generally refer to the reader as "he.") This figure of the reader of *Indian Opinion* plays a central role in the genesis of *Hind Swaraj,* which the chapter interprets as an account of reading and satyagraha, the one an analogue for the other. Since *Hind Swaraj* itself first appeared in the columns of *Indian Opinion,* the chapter analyzes it as a text woven out of the forms and genres of the paper.

The Conclusion returns to Gandhi's South African textual experiments by considering the link between his refusal to assert copyright and his invocation of the reader of *Indian Opinion* in the guise of a deterritorialized diasporic subject. The intersection of these two points—free textual circulation and the deterritorialized subject—highlights the radical nature of his ideas on sovereignty. These resided not in territorial space or abstract rights but in a zone of free textual circulation where readers anywhere claimed (and concretized) their own rights by copying and reproducing texts whether mentally or physically. The Conclusion takes these discussions of Gandhian textual theory into the present. How do his insights about summary and speed as a central feature of modernity and industrialized orders of information speak to conditions today?

Gandhi in South Africa

For readers unfamiliar with Gandhi's life in South Africa, I provide a brief overview.[44]

Born in 1869 in Porbander in Kathiawad, Gandhi grew up and received his initial education in this princely state. Between 1888 and 1891, he studied law in London and returned to India, where he practiced in Bombay and Rajkot. In 1893 he took on a brief from the firm of Dada Abdulla and Company in Natal, migrating like many Gujaratis before him out into the Indian Ocean world (rather than inland to Delhi or across the subcontinent to Calcutta). He was to spend the next twenty-one years in South Africa (with breaks to visit India in 1896 and 1901-1902 and trips to London to lobby on behalf of South African Indians in 1906 and 1909). In southern Africa, Gandhi encountered a patchwork of sovereignties: in the north, two independent Boer republics, the Transvaal and Orange Free State (OFS); further south, two British colonies, Natal and the Cape Colony. Both within their borders and beyond lay a series of independent or semi-independent African chiefdoms. In the wake of the Anglo-Boer War (1899-1902) and a British victory over the Boer republics, these units were drawn together into one country, the Union of South Africa (formed in 1910).

Indian communities were mainly spread across Natal and the Transvaal with a small grouping in the Cape and a handful in the OFS, a territory that sought to exclude all Indian immigrants. The Natal community was made up of an underclass of indentured laborers recruited from 1860 to 1911 to work on plantations, mines, and railways. A group of merchants (predominantly Gujarati Muslims) established itself in the wake of these laborers. About half of these laborers stayed on in South Africa after the expiry of their contracts, and their descendants came to comprise a "colonial-born" class whose numbers included a tiny strata of professionals, with some of whom Gandhi worked in the satyagraha campaigns. The Transvaal had virtually no Indian indentured labor, and its communities mainly consisted of merchants, traders, and small hawkers. By the early 1900s, there were about 150,000 Indians in South Africa: 133,000 in Natal; 10,000 in the Transvaal; 6,600 in the Cape; and 100 in the OFS.

From 1893 to 1903, Gandhi was primarily based in Durban and largely acted on behalf of merchant interests. Faced with discriminatory legislation and mounting racism, merchants had started to organize themselves even before Gandhi arrived. Himself a victim of these growing racist measures, Gandhi acted as a legal and political agent for merchants and was central to the formation of the Natal Indian Congress in 1894. The establishment of the IPP in 1898 formed part of this merchant-led political initiative by which they sought to portray themselves as worthy citizens of empire (while distancing themselves from the indentured classes and from Africans). This imperial emphasis remained critical throughout Gandhi's South African years, and he sought out occasions on which to demonstrate Indian loyalty to empire most famously through his mobilization of Indian ambulance brigades who worked on the British side during the Anglo-Boer War and the Zulu Rebellion in Natal in 1906.

After the Anglo-Boer War, Indian hopes were high that the new rulers would prove more accommodating than the old Boer regimes. Such hopes were soon dashed as anti-Indian legislation as vicious (if not more so) began to emerge from the British authorities in the Transvaal. This climate of crushed expectation promoted a feeling of urgency and activism of which the launch of *Indian Opinion* in 1903 forms one strand.

Establishing himself in Johannesburg in 1903, Gandhi was admitted to the Bar of the Transvaal Supreme Court. The cosmopolitan world of Johannesburg, a gold-mining center, brought into being in 1886 by migration from most quarters of the world, was an important influence in broadening Gandhi's thinking.

The satyagraha campaigns ran from 1906 to 1914. Their trajectory can be summarized as a drama with three acts: a vigorous opening, a period of retreat and near collapse, and a powerful conclusion.

The campaign began in earnest on September 11, 1906, when a mass meeting of Indians took an oath to defy proposed legislation requiring all Asians (Indians and Chinese) over the age of eight to take out identity certificates bearing their photographs and fingerprints. By November, only 545 out of a community of 7,000 had registered, and by January 1907, 2,000 resisters, including Gandhi, had been jailed for refusing to register.

Despite its initial successes, the campaign had internal weaknesses, aggravated by the toll of imprisonment. Fearing that their movement might collapse, Gandhi, Thambi Naidu, a leader of the Tamil community, and Leung Quinn, chairman of the Chinese Association, entered negotiations with Jan Smuts, a Boer leader and prominent cabinet member in the Transvaal government. An agreement emerged: Asians would register voluntarily in return for the law being repealed. This switch of direction occasioned considerable confusion and anger, causing one Pathan to assault Gandhi.

The compromise soon collapsed: the law remained on the statute books, and by August 1908 the campaign recommenced, with resisters who had registered voluntarily burning their certificates.

This resumption of the campaign can be seen as the "second act" of the drama characterized by difficulty and setback. The initial merchant support base of the movement evaporated. Facing imprisonment and the prospect of their assets being attached, most merchants abandoned satyagraha. Small hawkers, mainly Tamils, with less to lose took their place.

By 1909 the movement was under severe strain. The leadership was further weakened by jail sentences and desertions. Funds were desperately low. Deportations to India demoralized the spirit of resisters. Gandhi's leadership started to come under challenge from those keen to reach a settlement with the Transvaal authorities. At this low ebb, Gandhi undertook a deputation visit to London and on the return trip wrote *Hind Swaraj,* expressing his views on satyagraha and swaraj. These he had refined over the course of his campaigns, in the columns of *Indian Opinion,* and in his extensive reading during his three prison terms.

By 1910, with support from Hermann Kallenbach, Gandhi established Tolstoy Farm as a refuge from the remnant of the movement. Henry Polak, another of Gandhi's trusted lieutenants, undertook a fund-raising trip to India. Led by Ratan Tata, substantial donations from Indian industrialists came to the rescue of the cash-starved organization.

The third and final act unfolded between 1911 and 1914, when a series of circumstances enabled Gandhi to shift the scale of the movement to a national level. New immigration legislation from the Union of South Africa parliament threatened the capacity of Indians to move between provinces and hence permitted Gandhi to appeal to communities beyond just the Transvaal or Natal. In 1912, Gandhi's mentor, G. K. Gokhale, made a highly publicized tour of South Africa that raised Gandhi's standing considerably.

A judgment in March 1913 affecting the status of Hindu and Muslim marriages presented an opportunity to renew mobilization using the emotive call to defend the honor of Indian women. A final factor that dramatically extended the scale of satyagraha was the £3 tax. This annual payment applied to indentured workers who did not reindenture themselves at the end of their contracts and stayed on in South Africa rather than returning to India. Its abolition had long been a demand of "colonial-born" Indians and was finally taken up by Gandhi in September 1913. Widespread strikes on both sugar plantations and mines in Natal followed, with Gandhi leading some of these striking workers across the border to the Transvaal to court arrest.

Yet a further round of Smuts-Gandhi negotiations ensued, with Gandhi promising to suspend satyagraha in return for legislation that would recognize the goals of the movement. The Indian Relief Bill of 1914 was the law that eventuated. It abolished the £3 tax and addressed the issues around the legality of marriages. Other issues, like movement between provinces and limitations on Indian commerce, remained unresolved. Gandhi left South Africa on July 18, 1914, with the Relief Bill being hailed internationally as a triumph. Back in South Africa, Indian leaders remained critical of the partial nature of the settlement.

I

Printing Cultures in the
Indian Ocean World

At low tide from the beach in Porbandar in Gujarat, it is said one can just glimpse the tip of a shipwreck. Originally the SS *Khedive*, this vessel is the subject of a double legend. The first is that it carried Mohandas Gandhi on one of his voyages between Bombay and Durban. The second is that the ship sank with Gandhi's printing press on board.[1]

This is not the only report of a phantom printing press associated with Gandhi. More than a century earlier, another account had emerged, this time at the other end of the Indian Ocean, in Durban. Toward the end of 1896 Gandhi was headed for this port, on his way back from Bombay. Angry white mobs awaited his arrival, claiming he had traduced white settlers in the international media and was furthermore organizing an "Asiatic Invasion." Rumors circulated that Gandhi had a printing press and thirty compositors on board. In the imaginings of the white worker mob, the frontier of Asian migration moved with a printing press at its helm. The mob was organized into artisanal platoons: railway men, carpenters and joiners, store assistants, plasterers and bricklayers, saddlers and tailors, and printers. The pressmen on the port wharf carried a sign "Printers" that announced their view that the exclusive right to a press defined the white workingman's ideal of racial democracy (Fig. 5). According to the Natal media, the lynch mob intended first to attack Gandhi and then the printing press (*Natal Witness*, January 11, 1897).

In part this phenomenon of the phantom printing press can be easily explained. In the Durban case, it had been well known for some years that Gandhi wished to acquire a press in order to start a newspaper that could speak for the interests of Indian merchants in Natal.[2] The alarm that much of white Natal society felt over this idea expressed itself in the rumor of the imaginary press. White artisans were equally perturbed by the prospect of cheap compositors coming to undercut the market.

A few years later, in 1898, Gandhi did indeed play a part in purchasing a press, which was to print his newspaper *Indian Opinion*. When he returned to India in 1914, newspapers continued to be a key component of his noncooperation movements. Gandhi, the newspaper, and the printing press were closely associated, and hence the memory of Gandhi's printing press persisting in legend is understandable. Indeed, Gandhi's press has become something of a minor icon in his life story. A chapter of his autobiography discusses his printing press in South Africa, while the presses he used in India and South Africa feature in exhibitions and museums.[3]

Yet pertinent as these explanations are, they do not clarify one important aspect of these stories: both are accounts of a printing press at sea. Furthermore, both feature a press unable to make landfall in the Indian Ocean. In Durban, the guardians of the global color line keep the press at bay. In the Porbandar case, the press comprises part of Gandhi's South African baggage that cannot endure the Indian Ocean crossing. Unable to survive in South

Africa or transplant itself to India, its utopian potential drowns in the Indian Ocean.

The Indian Ocean has long been a site for utopian imaginings: peaceful cosmopolitan trade, *mare liberum,* island utopia, pirate republics, and, more recently, a series of transoceanic dreams, whether of theosophy or Greater India, followed by ideals of nonalignment and Afro-Asian solidarity.[4] In the reckoning of some, the Indian Ocean's ancient history of transoceanic trade enables utopian thinking, an argument made in Amitav Ghosh's *In an Antique Land.*[5] Ghosh himself has given the book a utopian gloss, describing it as a narrative of nonalignment, which explores again the ancient transoceanic exchanges of the Indian Ocean interrupted by colonial rule.[6]

This portfolio of Indian Ocean utopianisms provides an appropriate setting for the story of the International Printing Press (IPP), a modest jobbing press that became a central protagonist in the satyagraha saga. A "colonial-born" press, it belonged to a class of similar tiny jobbing operations that dotted the Indian Ocean world and that likewise printed material on behalf of utopian causes. Like the IPP, these Indian Ocean presses took shape in multiply diasporic environments. Taken together, they constitute an important but little-explored strand in global histories of print. This chapter traces the outlines of these Indian Ocean printing cultures as a framework for the biography of the IPP that follows in the next chapter.

Diasporic Printing

There are currently rich legacies of scholarship on transnational histories of print. These include work on imperial print culture and postcolonial book history, as well as studies of newspapers and periodicals in empire.[7] Yet these studies tend to operate in circumscribed areas: the transatlantic book trade; print and printers in the "white dominions"; and media in British India.[8] Studies of transnational religious publishing are likewise balkanized and tend to follow the watersheds of particular faiths. Accounts of Christian mission publishing in sub-Saharan Africa, for example, pay little attention to Indian diasporic print ventures or African commercial concerns, creating the impression that all printing in this region is of European Christian evangelical provenance.[9]

To complicate such views, one need only consider the printing sector in any of the major cities in southern or East Africa in which several diasporic groups intersected. In these settings, Muslim printers from Bombay; Africans tutored in Protestant evangelical presses; Indians (and Britons) trained in mission, state-run, or commercial printing concerns in the subcontinent; and British printers as well as print workers from diasporic locales like Mauritius converged.[10] These "journeymen" shaped multilingual, multiracial, and multireligious printing ventures, working with and against each other.

Generally financially tenuous, these small printing operations produced material on behalf of (or at times came into bring through) organizations espousing grand transoceanic schemes, whether Hindu reformist, Sikh transnationalist, African nationalist, pan-Islamic, or white laborist. Central to these repertoires of transnational communication were streams of print culture: blizzards of periodicals, pamphlets, leaflets, and tracts setting out the programs and principles of their makers wafted around the port cities of the Indian Ocean. These flows of printed matter created maritime markets of faith, ideology, and information (to adapt Nile Green's terminology).[11] Examining the case of Bombay's variegated

Muslim traditions, Nile Green demonstrates how the city's vernacular publishing industry produced a stream of demotic religious material—hagiographies, miracle narratives, cures, and talismans carried by holy men both further inland and outward to sea, in some cases to new, underdeveloped markets like Durban.

In obtaining this transnational media, diasporic organizations relied on the print cosmopolises of the Indian Ocean—Bombay and Cairo—but also created new presses and smaller centers of publication in places like Durban, Beira, Zanzibar, Mombasa, Nairobi, and Port Louis. These diasporic presses were wrought from different printing traditions carried by migrant groups from across the Indian Ocean and beyond. Two major streams emanated from Britain and India.

Printing Frontiers: Greater India Meets Greater Britain

British and Indian printers were global figures with an international profile and reach. By the later nineteenth century, major printing centers in both places had long been extruding print workers into empire. Unimpeded by immigration restrictions, British printers moved freely across empire, and especially the dominions where workers could refine a sense of themselves as "white."[12] These laboring men were less print capitalists than what we might call "print laborists"—men who attempted to define printing as part of white racial privilege. The print capitalists, proprietors, and master printers for whom they worked shared their racial ideologies if not their laborist proclivities.

More limited in their capacity to move, Indian printers did nonetheless make their presence felt across much of empire. They were drawn from a sizable Indian printing industry, which by the late nineteenth century had developed major nodes in Bombay, Calcutta, and Madras, as well as in Lucknow, Kanpur, Allahabad, and Lahore.[13] Fueled from the 1820s by a mixture of colonial-state printing, Protestant and Catholic mission vernacular presses, the spread of appropriate technology (the portable iron press, lithography), a paper industry, and accelerating communications networks, the printing sector was substantial and differentiated.[14]

In addition to large European-run concerns, there were sizable Indian-owned operations. As Ulrike Stark's wonderful history has shown, the Nawal Kishore Press in Lucknow employed 350 hand presses, a "considerable number" of steam presses, and 900 employees producing material in a range of languages.[15] By 1909, Madras was home to 120 printing concerns and sixteen factories for manufacturing printing presses.[16] Large Indian-owned Madras printer-publishers in the fashionable areas of the city like Ananda Press, Andhra Patrika Press, and Perumall Chetty & Sons employed several hundred employees each.[17] Beneath these strata of elite printing lay a sprawling world of smaller ventures, some comprising one- or two-man operations. As Anindita Ghosh shows for Calcutta, these printers huddled together in the narrow lanes of north Calcutta, producing a steady stream of popular almanacs, pamphlets, and images.[18]

Unlike the dominions, British India did not have an apprenticeship system, and forms of training varied, from learning on the job, to instruction in state-run colleges like the Madras Technical Institute for Printers or the Roorkee Engineering College, to tuition in prisons where substantial quantities of government printing were done.[19] Lacking the tight solidarities of apprenticeship, printing in India acquired varied meanings. In some cases printing was closely associated with caste, where particular artisanal groups like smiths and artisans in Calcutta had moved into the trade.[20] In other cases religion and caste shaped the meaning of printing, where lower caste or dalit groups were drawn into Christian mission presses.[21] Among proprietors, printing acquired a strong aura of reform and progress, with the figure of the printer-editor-proprietor an almost stock character of reformist movements across the subcontinent.[22]

Greater Indian Printing Diasporas

While there is a rich body of work on print traditions within India itself, the question of the printing diasporas it generated is something about which we know less. But move Indian printers certainly did. In some cases they followed their own linguistic communities into the diaspora, pursuing established trajectories like those from western India to East Africa, Mauritius, and southern Africa, the latter two destinations for south Indian artisans as well.[23] These printers initially worked in existing concerns or in cases where they had capital, set up their own presses, usually buying secondhand equipment and importing type from India or, more rarely, setting up lithographic operations. The earliest Tamil printing on Mauritius dates to 1843, followed by lithographed Gujarati and Urdu in 1883, and then by Urdu, Gujarati, and Hindi in letterpress.[24] Printers also petitioned to bring out fellow workers from their hometowns, requests that were not always approved given the imperial restrictions on Asian migration. Once established in the diaspora, these printing concerns became nodes of expertise, training, and onward migration: Mauritius hence became a source of recruitment for Durban printers.[25]

Religious networks provided another avenue of movement for pressmen. As Nile Green's study of Muslim printing in Bombay demonstrates, men and machinery from the cosmopolis of the Indian Ocean traveled outward to destinations in East Africa, South Africa, Southeast Asia, and central Asia.[26] While little is known of mobility among Christian mission-trained printers, these may well have migrated too or, in some cases, moved via Indian Ocean mission networks.[27]

There is also some evidence that printers moved on a semivoluntary or indentured basis and were "requisitioned" by colonial agencies as a form of skilled indentured labor. One such case occurred in Natal in 1896, when the Immigration Trust Board approved the importation of a dozen indentured Indian artisans, which by some accounts included compositors. The Natal Typographical Union objected strenuously and the contracts were canceled. Rumors of skilled labor coming from India nonetheless continued to circulate among white artisans and helped fuel the climate that fed into the near lynching of Gandhi, discussed earlier.[28] There is also evidence from Australia of Anglo-Indian printers being brought over from Madras but, again, such schemes ran into opposition from organized white labor.[29]

Another strand of printing expertise from India consisted of British printers who had either migrated to, or had been trained in India or had worked there before moving on to other colonial destinations.[30] The figure of the journalist is an important analogue in this regard, and Gandhi was to encounter several India-trained journalists in South Africa keen to help or hinder his project.[31]

Printing Encounters

The encounter of these two printing frontiers—British print capitalism *and* print laborism, on the one hand, and Indian diasporic print formations, on the other—produced its own vocabularies and terminology. The phrase "colonial and Indian printing" divided the empire into three printing zones, one implicitly the metropolis, the other two the dominions and British India.[32] White printers in Natal, and probably in other dominions, used another term, "oriental printing," to designate Indian-owned presses.[33]

These interactions produced their own stock of stereotypes, including the ideal of the white manly printer and his opposite, the "effeminate" Indian printer, who lacks the discipline of apprenticeship. In an image circulated among white dominion printers, the Indian compositor invariably makes gratuitous "oriental" swirls and flourishes while typesetting, bows to the case, and talks or sings on the job.[34] Another "shortcoming" lay in page makeup and design. Indian printers favored an overdecorative style apparent in title pages set with types intended for handbills and advertisements rather than books and liked "inappropriate" "old-fashioned fancy types."[35] There were also "Indian methods" of doing things in the print shop that white printers deprecated: unnecessary movement of hand to galley that so appalled one observer that he measured the wastage (eight hours of "oriental flourishes" amounted to three thousand gratuitous yards of movement) or setting type without the aid of a setting rule. Defective "country-made" equipment caused further inefficiencies—North Indian frames were "wrong," since the slope for the upper case was the same as for the lower, resulting in "much waste of the compositor's time, as his hand travels a longer distance than is necessary."[36] Together these flaws demonstrated all the "vices" of the "foreigner"—a term used by printers to designate those who had not done apprenticeships.[37]

Other Encounters

The encounter of these two printing frontiers unfolded in worlds that included other printing traditions. In a port city like Durban, for example, different modes of printing congregated: Protestant mission, Indian diaspora, maritime Muslim, colonial state, British capitalist, and laborist. In addition, African printers and print proprietors were present. Christian mission secondary schools and training institutes with an industrial and vocational emphasis had long trained African students in different aspects of printing. Generally debarred by law or prejudicial custom from white-owned concerns except as manual laborers, skilled African printers were limited to seeking employment in Christian mission presses that dealt with African language material or with the minute number of African-owned presses.[38] These latter operations were generally established by members of the African elite seeking to establish presses free from white mission control.[39] Famously, one such press belonged to John Dube at Ohlange Institute, next door to Phoenix.

In later years, a further outlet for African printers would be Indian-owned concerns. In Kenya, Indian traders and entrepreneurs not only facilitated the spread of trade goods and finance but were also decisive in initiating and sustaining a culture of printing and print media, particularly newspapers, across much of East Africa. These presses became sites for African-Indian cooperation and competition: Africans bought equipment from Indians; Africans worked in Indian-owned presses; Indian printers assisted

African nationalist causes; and Africans sought to establish presses free of Indian control.[40] One important example of such African-Indian interaction from the early 1920s involved the early Kenyan anticolonial figure Harry Thuku and M. A. Desai, the editor of *East African Chronicle*. As Sana Aiyar explains, Thuku took Gandhi as one of his political models and received support from Desai and the East African Indian National Congress.[41] The presses of the *Chronicle* printed Thuku's pamphlet, causing sufficient alarm in official circles for the Chief Native Commissioner to order a search of the premises.

In South Africa, this African-Indian interaction around the printing press began from the 1920s, establishing important new intellectual conjunctures. As Thembisa Waetjen and Goolam Vahed have shown, in 1937 Chamberlain Nakasa, the typesetter at the Union printing press, which printed *Indian Views*, went on to start the short-lived monthly *New Outlook*, which drew together African and Indian opinion.[42] In the post-Gandhian era at Phoenix, as Uma Dhupelia-Mesthrie has shown, Alpha Ngcobo joined the press as a salaried worker in 1921 and continued for three decades.[43] In the period examined in this book, however, such interaction was largely unknown, with African and Indian printing working in isolation from each other, each as we shall see pursuing separate projects of the ethnic regeneration of Africa and India.

"Colonial-Born" Presses

These congeries of printing traditions spawned a series of "colonial-born" presses up and down the East African littoral in places like Durban, Beira, Dar-es-Salaam, Zanzibar, Mombasa, and Nairobi.[44]

We currently know little about these presses beyond a brief glimpse of them on letterheads, in newspaper advertisements, in applications to the colonial state to register the paper, in a censor's report, or via the small body of academic discussion of the newspapers and their presses.[45]

Drawing together this scattered information we can attempt to draw up a rough typology of such presses. To do so, we start with a description of two presses in Kenya drawn from Zarina Patel's work.[46] The first is the Khalsa Press, set up in the 1920s by Sudh Singh, father of the great Kenyan anticolonial figure Makhan Singh. A solo operation run out of a one-room home in Nairobi, the press did work in Gurumukhi and served the Sikh community. Singh kept his printing material under the bed, and even when orders were slow he ran the press to create the impression that business was flourishing. The press did job printing as well as pamphlets, poetry, and religious texts for two Sikh temple associations. Despite being so tiny, the press had an international reach via Sikh networks; it was advertised in a magazine in the Punjab and had links to the transnational Ghadar Party that sought the armed overthrow of British rule in India. In subsequent years the press expanded and opened in larger premises, offering printing in Gurumukhi, Gujarati, Urdu, Hindi, English, and African languages.[47]

If the Khalsa Press represented the bottom end of "colonial-born" presses, our second example, the African Standard Printing Works, in Mombasa, represents the top. Established in 1902 by A. M. Jivanjee, the larger-than-life entrepreneur, empire builder, and "Grand Old Man of Kenya," the press included an imported high-speed press and associated equipment and a staff of about a dozen to print the English-language paper *African Standard, Mombasa Times and Uganda Argus* (to use its full title). The paper had been set up to counter the white settler press and give expression to an ideal of British Indians as citizens of empire.[48]

Judged by British or Indian standards, Jivanjee's press was small, Singh's minute. Yet despite their limited size, both harbored substantial ambitions, whether related to Sikh transnationalism or ideals of imperial citizenship. These were small operations with large aspirations, a combination that conferred a contradictory character on these presses.

On the small end of the equation, most "colonial-born" presses were often family run and economically tenuous, and in this latter respect, they resembled the numerous and equally tenuous white-run jobbing presses across much of empire. Across the board they struggled to break even from job printing alone, and all had to supplement their business with other activities—printing newspapers and pamphlets and selling stationery, books, or patent medicines.[49]

With regard to the larger aspirations of these small presses, these were articulated in periodicals and newspapers with ambitious messages of social reform and "progress." The Ottoman Printing Press in Durban announced its main business as printing *Al-Islam*, "First Mohamedan Weekly in South Africa" (established in 1907). The IPP advertised itself as the publisher of *Indian Opinion*, "In the Interests of British Indians in South Africa."[50] These periodicals and their causes defined the press, often bringing it into being and conferring their title on it as in the case of the African Standard Printing Works.

In this world of ideological printing, the figure of the editor-proprietor-publisher constituted an important character in the transnational drama of social reform. Yet this job description is not something predetermined or agreed upon, especially in a world

of limited resources, elastic divisions of labor, and lack of special-
ization. "Colonial-born" presses had to rely on extemporization
and ad hockery. In the case of *Indian Opinion,* its editors were to include a commercial agent (with some freelance journalistic experience in London) (M. H. Nazar), an electrical contractor (Herbert
Kitchin), an administrative manager from London who began his
journalistic career in Johannesburg (Polak), and a Baptist minister
(Joseph Doke).[51] As Polak rightly observed, it was a world of the
"educated amateur" that in certain respects resembled a more com-
plicated version of the Victorian environment of "old journalism,"
with its gentleman editors and commitment to improvement.[52]

To complicate matters further, the job of editor was seldom anyone's sole occupation. P. S. Aiyar, editor of the Natal paper *African
Chronicle* (started in 1908), doubled up as a translator and interpreter
for the Supreme Court.[53] These editors resembled many of their
counterparts in India, whom Stark has characterized as taking on
the roles of "entrepreneur, publicist, literary patron, philanthropist,
disseminator of knowledge and educator."[54] Like many colonial intellectual figures, these editors had to be men of many parts: law-
yers, journalists, editors, social workers, historians, and politicians.

This multiplicity of roles meant that editors and proprietors were
pulled in different directions, as Jim Brennan's analysis of two Tanganyika newspapers, the *Tanganyika Opinion* (1923–1955) and *Tanganyika Herald* (1929–1962), demonstrates.[55] Both fragile operations, the
papers and their presses (Kanti Press and Herald Printing Works)
bent to the prevailing political winds, supporting an incongruous
range of ideals: imperial citizenship, Hindu reformism, communism,
and anticommunism.

The manifold contradictions facing "colonial-born" presses were
further exacerbated by the social mission of the papers, which cut

across the need to turn a profit. These conflicting demands meant that these "colonial-born" presses had to be both entrepreneurial and philanthropic, pursuing projects of social reform while mobilizing networks of charitable support from merchant donors and, in some cases, from readers. However, in diasporic contexts, philanthropic opportunities were somewhat limited, forcing newspapers to be more "modern" than some of their counterparts in India that publicly maintained differential subscription rates with patrons giving royally, thus enabling some newspapers to keep afloat with as few as 150 subscribers.[56] Diasporic papers had set subscription rates and had to find a larger audience. *Indian Opinion*'s circulation figures ranged from 800 to 3,500, with an average circulation of 1,200 to 1,500 (with a secondary readership of borrowers and illiterates around these core subscribers). The lower-end figure is often seen as small, especially for a Gujarati community of 20,000.[57] Yet when compared to some publications on the subcontinent, these figures appear more substantial. Importantly, in keeping with being "modern," very few "colonial-born" presses used lithography, preferring letterpress as the more "civilized" and progressive alternative.

A second modern feature of these presses arose from their diversity, which made them more cosmopolitan than their British and Indian counterparts. The former were generally monolingual and at times monoracial; the latter were still largely dominated by caste division, with Brahmins doing the writing and intellectual work and lower castes undertaking the physical printing. Diasporic presses, by contrast, drew together personnel made up of different castes, languages, religions, and races (but seldom different genders). The work always involved a minimum of two languages and multiscript compositing. Translation was hence central to the workplace and the labor process itself.

These "colonial-born" printing presses unsettle any self-evident meaning that we may ascribe to printing and publishing. They drew together different printing traditions from across the imperial world and enabled these to articulate with each other. In so doing, these small presses established new pathways of publicness. Both in the physical world of the shop floor and in the printed matter they produced, these presses connected groups on as yet unimagined latitudes among different colonies. The story of the IPP, which began life as just another obscure "colonial-born" press, illustrates the world-changing potential inherent in such alignments.

2

Gandhi's Printing Press

A Biography

The International Printing Press (IPP) came into the world in considerable style on the evening of November 29, 1898. An opening ceremony attended by a crowd of nearly a hundred inaugurated the press at its "commodious premises" at 113 Grey Street, next door to the Natal Indian Congress (NIC) Hall. A report in the *Natal Mercury* (November 30, 1898) gave some background: "The necessity of Oriental printing works has been felt for a long time past by the Indian community in Natal, and especially by merchants, priests, and teachers in Durban and other centres of population in the Colony."

The proprietor, Viyavarik Madanjit, a Bombay schoolteacher, welcomed the guests and invited Abdul Kadir—a well-known merchant who had provided the press with rent-free premises—to preside. On taking the chair, Kadir requested the Congress organist to play "God Save the Queen." Speeches followed in Gujarati, with Gandhi translating (*Natal Mercury*, November 30, 1898).

Madanjit sketched the history of the press. Three years earlier Gandhi had asked him to set up a small press with the support of leading merchants.[1] The idea came to naught, but after some time Gandhi had again put the scheme to him, this time with a utopian touch. There would be no merchant support, and Madanjit would depend on his own means. Workers would not be paid a salary but would labor for a share of the profits. The press could benefit the community in that "it would enable our young men if they so cared, to learn the art of printing as well as, possibly, to earn their living." Gandhi was keen to get the press going so that it could print material in Indian language scripts (*Natal Mercury,* November 30, 1898).

The scheme appealed to Madanjit's "patriotic instincts," and with the assistance of a Pietermaritzburg printer he purchased a secondhand printing plant, which included 1,000 pounds of English type. Gujarati and Hindi type had been ordered from India, with Tamil to follow. Madanjit knew nothing about printing and so had engaged Mr. Ratanshi, who had worked in a "large printing establishment in India, making him a co-worker and sharer in the profits." The Pietermaritzburg printer who had joined the concern then resigned, and Ratanshi took charge. A bookbinder, Mr. A. Guanarathinam Pillay joined the press and expressed his willingness to learn printing on the job (*Natal Mercury,* November 30, 1898).

Madanjit had hoped that "our young men" would join the press on "a co-operative basis," but this scheme, which in effect amounted to an unpaid apprenticeship, proved unattractive. He appealed to the audience to recommend two Tamil youths fluent in Tamil and English, as well as two further apprentices (*Natal Mercury,* November 30, 1898).

Madanjit continued: "The undertaking is at present a speculation. We have all believed a printing press to be a necessity amongst us. No civilised community can do without it." He urged the community to play its part by bringing their circulars, delivery books, invitation cards, sporting notices, and programs to be printed. "The press is not mine alone—it is yours also. I venture to think that the Indian community is as much interested in it as Mr. Ratanshi and myself." Abdul Kadir (whose subsidy of the press's premises belied Madanjit's more utopian pronouncements) declared the press open and urged the audience to have their circulars "in Oriental languages" done at the IPP (*Natal Mercury*, November 30, 1898).

Gandhi read letters of support from Indians in Pietermaritzburg, Umzinto, and other centers in Natal. The press was then declared open. In the words of the newspaper report: "A Brahman priest and teacher from Verulam [a settlement north of Durban]" apostrophized the press with some Gujarati verses. "He recited the praises of the art of printing and the manifold blessings that follow its development." He thanked "Queen Victoria for the freedom which enabled them to obtain the privileges and blessings accruing from printing."[2]

Speaking in English, Canon Booth, head of the Anglican Indian Mission in Natal, said that a press "in any community marked a distinct step in their progress." English people in Natal were at times alarmed "when they heard about the vernacular press, because they knew that insignificant rags of newspapers printed in the vernacular in India had done much mischief." The IPP, however, was different, as it stood under the loyal patronage of the NIC, a comment that evoked "loud cheers." Printed English copies of Madanjit's speech were made available (*Natal Mercury*, November 30, 1898).

The IPP at Grey Street

The opening ceremony announced the press as an unusual institution, with all the "in-between" characteristics of a "colonial-born" press. Situated among a range of media and printing traditions, the IPP stood between different languages and scripts; between print and performance; between profit, community service, and merchant patronage; and between grand objectives (advancing civilization, profit sharing, imperial loyalty) and lowly job printing. Geographically the press was located between different continents and port cities, combining expertise from Bombay (whence Ratanshi almost certainly came), secondhand equipment from Durban's hinterland, and bookbinding skills from Madras. This mix relied on both professional expertise and Madanjit's amateur enthusiasm. As an institution working with Indian languages, the IPP was set alongside "insubordinate" presses on the subcontinent in opposition to which it had to be loyal and exemplary. A heavy mantle had fallen on a modest jobbing press, its task none less than combining "oriental" and "occidental" printing, East and West.

The IPP tackled its onerous mission with limited equipment that consisted of two hand-operated presses: the first, an Albion Press (the most widely used press in empire), and the second, a platen jobber, as well as English type.[3] In addition to the Hindi, Gujarati, and Tamil type mentioned earlier, cases of Urdu and Marathi type from India followed.[4]

The press undertook general job printing. It printed the monthly magazine of the theosophical society and a short-lived newspaper called the *Volunteer,* as well as booklets and pamphlets.[5] From June 1903, the IPP added the printing of *Indian Opinion* to its schedule. As we have seen, the press advertised its ability to do work in ten languages—English, Gujarati, Tamil, Hindi, Urdu, Hebrew, Marathi, Sanskrit, Zulu, and Dutch (with seven scripts)—the first four in all likelihood generating most of the work (advertisement on cover, June 11, 1903).

Its initial personnel were multilingual and drawn from across the Indian Ocean and beyond. The earliest employees—the anonymous printer from Pietermaritzburg, Ratanshi, and the bookbinder, Pillay—appear not to have lasted long. In Pyarelal's account, the longer-term employees consisted of the foremen, Mr. Oliver from Mauritius, who oversaw a staff of typesetters and machine men. The English compositors included a French-speaking Mauritian, a man from St. Helena, and, in the parlance of the day, a "Cape coloured" Mr. Mannering. The Gujarati typesetters were Kababhai and Virji Damodar Mehta, with Virji doing Hindi as well. Moothoo, a "colonial-born" Indian, undertook the Tamil composing, while Raju Govindswamy ("Mr. Sam," as he was known) was in charge of machines and binding.[6] Also "colonial-born," Govindswamy had started his career as a "kitchen boy" in Umkomaas and had then moved on to being a messenger on the railways, before becoming an assistant at thirty shillings per month in a printer's firm where he was recruited by Mandanjit.[7] In Albert West's account of the press, there were also "several young Indian printer's assistants."[8] The idea of unwaged work died quickly, and the enterprise paid salaries that were generous by Natal standards, ranging from £8 to £18 per month in an industry whose average wage was £12.[9] One employee was a member of the South African Typographical Union (SATU) and had had to seek permission to work for an Indian.[10] The other workers were not unionized.

We have little detailed insight into the workings of the IPP in its early years. However, at the inception of *Indian Opinion*, the correspondence between the first editor, M. H. Nazar, and Gandhi, who was based in Johannesburg, provides some angles on this topic.[11]

Initially a small jobbing press, the IPP suffered from a lack of skill and experience. Madanjit had no printing experience at all and seemed reluctant to acquire any. Nazar described his attitude to his responsibilities as "elastic," and he was often absent. The routines in the IPP were muddled. "You have no idea of the time I have to *waste* in the Press," he wrote.[12]

Compounding these problems was the difficulty of doing job and newspaper printing in multiple languages. Type in Indian languages was in short supply, and as Nazar explained in relation to the compositing of *Indian Opinion:* "Hindi and Tamil must be set, printed and distributed, and re-set if you want 6 or 7 columns of each." The Gujarati case had a shortfall of the character "a," and on occasion Virji, the typesetter, asked Nazar to avoid words containing this letter.[13] Tamil and Hindi were to be dropped after two and a half years (although these two languages were briefly resumed in December 1913, but only for four months). Most Gandhi scholars typify the duration of the four-language policy as short, but when seen from the point of view of the shop floor, it must have seemed interminable. Keeping the four-language policy going for that length of time was hence no small achievement. Indeed, the difficulties of the undertaking were to shape Gandhi's orthographical thinking and in subsequent years turned him into an apostle of Devanagari—the script that he argued should be used for all Sanskrit-based languages.[14]

Like much else in the IPP, the business of how to work in so many languages had to be made up on an ad hoc basis, as staff grappled with how to negotiate translation in the labor process. Copy appears to have been produced in English and then passed on to outside translators, while Nazar did some of the translation into Gujarati. The process was cumbersome and added two to three days to the weekly schedule. Nazar complained that the quality of translation was poor, and when he could he reworked it.

English material for the Hindi and Tamil translators had first to be simplified, as their English was uncertain. Translators insisted they could only do the work at night, possibly because they were being asked to do it on a voluntary or semivoluntary basis in keeping with the ethic of the paper, which placed service above profit.[15] Since Nazar spoke no Tamil, the business of linguistic quality control probably fell on the typesetter, Moothoo. The Hindi, Gujarati, and Tamil typesetters worked at a snail's pace, producing one and a half columns per day, an amount that a highly experienced compositor, drilled in the military discipline of apprenticeship, might have done in a third of the time.[16]

On a daily basis, Nazar had to invent his role as "editor," a position he undertook without a salary or a clear job description. As the man in the middle, he had to try and make the operation work, acting variously as a translator, rewriter, shop floor manager, linguistic quality controller, proofreader, and public figure. He also had to manage an enterprise that was distributed between Durban and Johannesburg, 365 miles inland, where Gandhi was mostly based. While never the editor of *Indian Opinion,* Gandhi was the senior statesman of the operation and a key writer whose work, despite Nazar's pleas, did not always arrive on time. As the man on site, Nazar also had to manage labor relations, a tricky task in an environment where deadlines routinely ran over schedule and workers were called upon to do considerable amounts of overtime. Nazar wrote: "They can't work till midnight for 3 consecutive days, under high pressure week after week. . . . If this continues, grumbling will end in a strike."[17] To Nazar fell the exacting task of negotiating the interface of public service and the nitty-gritty of printing and journalism.

As an instrument of social reform, the IPP was embedded in a series of broader public service networks on which the press could draw. The first of these was the NIC, which provided patronage and support. The IPP's senior personnel were all affiliates—Gandhi a founder, Nazar the joint secretary, and Madanjit a member.[18] A second network was the Indian Volunteer Ambulance Corps, which Gandhi had constituted to serve on the British side in the Anglo-Boer War (South African War) of 1899–1902. The Corps had IPP links: its members included Gandhi, Madanjit, and Nazar, as well as a future editor of *Indian Opinion,* Herbert Kitchin, while the indentured workers who made up its rank and file provided much of the labor when the press moved from Durban to Phoenix.[19]

White Printers

By the turn of the century, Durban was home to fourteen presses, all of them white-owned except for the IPP (Fig. 2) (there were also Christian mission presses outside the city, discussed later). The IPP was located on Grey Street, the *Indian Opinion* office on Mercury Lane, in the heart of the city's small printing district. Grey Street fell into what had emerged informally as the Indian area, while Mercury Lane was located in the "white" foreshore of the city, although there was some porousness between these areas. This flexibility was, however, rapidly disappearing as the Natal colonial state began implementing ever-more-rigorous forms of municipal segregation.[20]

With regard to these monoglot print shops in the white area, these would generally have been staffed by white men except for menial positions, which would have been occupied by Indians and/or

Africans. The white printers were united by strong bonds of solidarity. Most printers had done apprenticeships in Britain, parts of the empire or Natal, and they belonged to the well-organized and vocal SATU.[21] These close-knit solidarities (including a strong culture of drinking) promoted a marked sense of racial protectionism and white laborism. Like the IPP, these presses were embedded in networks outside the shop floor, one of which ironically mirrored the Ambulance Corps.

One key institution of colonial Natal society was the volunteer military regiment, which supplemented the small imperial standing army and promoted a widespread ethic of militarized masculinity, as Robert Morrell has demonstrated.[22] At times, men who worked together signed up for these regiments, as did a group of typographers who joined the Frontier Light Horse in 1879 in the lead-up to the Anglo-Zulu War.[23] In the "Asiatic Invasion" crisis of 1896, this cross between the volunteer regiment and the workplace became apparent: as mentioned earlier, lynch mobs were organized into divisions according to their artisanal professions, including printing (*Natal Witness*, January 15, 1897) (Fig. 5).

Yet the relationship between Gandhi and white printers was not solely adversarial, and one of the key players in the IPP was to be Albert West, a trained printer from Lincolnshire, whom Gandhi had met through vegetarian circles in Johannesburg. As problems at the IPP mounted, Gandhi asked West, who ran his own printing concern in Johannesburg, to take over the press. West agreed and went down to Durban, where he reported deep financial and management problems. Madanjit had been wanting to return to India for some time, and so he ceded the press to Gandhi to settle the debts he had accrued and then duly left. The IPP had, in effect, been kept afloat by large subsidies from Gandhi, a situation that could not continue indefinitely. In early October 1904 Gandhi caught the overnight train down to Durban to sort out the affairs of the press, and on the journey he famously read Ruskin's *Unto This Last*, which inspired him to take the radical departure of setting up his first ashram, Phoenix, fourteen miles north of Durban.[24]

Phoenix's Hinterland

While some accounts portray Phoenix as a remote and isolated outpost, it formed part of a crowded ideological hinterland where African American–inspired Zulu nationalists, Protestant evangelists, Arya Samajists, and Bombay Muslims intersected. The IPP comprised but one element in this mosaic of projects in which ideas of self-help, improvement, and uplift from both the black Atlantic and the Indian Ocean encountered each other. Yet while these projects spoke a shared idiom of reform and progress, they maintained an arm's length attitude to each other, subscribing in different ways to Booker T. Washington–style notions of segregational self-sufficiency.

These encounters were most prominently embodied in two neighboring institutions, Phoenix and Ohlange, the latter established by John Langalibalele Dube in July 1901 (Fig. 4). Often known as the Tuskegee of South Africa, Ohlange constituted a unique cosmopolitan experiment that threaded together Dube's formation in Zulu nationalism, Christianity, African nationalism, and black internationalism that was shaped by his studies in the United States, and his immersion in different strands of African American thought. The coeducational school attracted students from across southern Africa who spoke a range of languages and staff from the subcontinent and further afield, including the West Indies. It was a unique experiment, "an all-African school, free of both mission and state oversight," as Heather Hughes notes in her excellent biography of Dube.[25] The school formed part of Dube's larger empire in which he towered as a churchman, politician, educationalist, editor, and writer. His fame today resides largely in his position as the first president of the African National Congress, an office he occupied from 1912 to 1917.

Our interest in Dube is somewhat more tangential and concerns an iron hand press that from June 1903 began working on the premises at Ohlange.[26] A gift from Dube's U.S. supporters, the press played a part in the printing of his newspaper, *Ilanga lase Natal: Ipepa la Bantu* (Sun of Natal: The Black People's Paper). (The hand press was probably used to pull proofs, with mechanized machines doing the actual printing.) The print shop formed part of Ohlange's industrial section, where it was run by students under the management of a West Indian teacher, Reynolds Scott. The trilingual paper (in Zulu, English, and Sotho) was started in April 1903, with its initial editions having been printed at the IPP.[27]

There is only one faint photograph of the press (*Ilanga lase Natal*, December 22, 1933), and hence its exact make is uncertain, but it could possibly be a Washington press very similar in design to the Albion at the IPP.[28] When the IPP moved to Phoenix in December 1904, these nearly twin presses clattered away at the heart of two institutions whose projects resembled each other in form but not content. The founding motto of Ohlange, "to teach the hand to work, the brain to understand, and the heart to serve," almost exactly captures the spirit and intention of Phoenix.[29] Both communities were centers of self-help, religious reform, and ethnic regeneration in pursuit of which they printed newspapers, pamphlets, and books.

Yet the leaders of these two remarkable communities kept their distance and met rarely (in part because Gandhi was largely based in Johannesburg, and a fund-raising trip took Dube to the United States from 1904 to 1905). Both expounded different versions of "race pride," with Dube involved in redeeming "Africa" and Gandhi in nurturing "India" (Phoenix, in Gandhi's words, being "a nursery for producing the right men and the right Indians" [CW 10: 60]).

These separatist sentiments emerged as much from colonial triage above as from popular segregationist sentiments below. As Heather Hughes has shown, in the area around Phoenix, African tenant farmers and Indian market gardeners jostled for space on limited marginal land.[30] While these communities exchanged experiences, especially in the realm of healing and religion (a small group of Indians joined Isaiah Shembe's Nazarite church, a stone's throw away from Phoenix and Ohlange), there were also major fault lines between these groups.[31] Africans' rent was reckoned by hut, and they paid more than Indians, who were charged by plot. Anti-Indian sentiments were also propagated by U.S. Protestant missionaries at neighboring Inanda, keen to protect "their" African flock from adverse "foreign" influences.[32]

These twin printing projects at Ohlange and Phoenix were not the only ones in the Durban vicinity. Thirty miles further south lay Adams College (under the auspices of the American Board of Commissioners for Foreign Missions), which housed a press. (Dube had in fact attended Adams; however, he did not learn printing here but acquired the skill later in the United States.) Yet another press was attached to Mariannhill, a Trappist community that Gandhi had visited and admired enough to write about in 1898 (Fig. 3). As Michael Green has shown, the Trappists were unlikely evangelists. Dedicated to solitude, meditation, and withdrawal from the world, their proselytization was pursued through the slow and silent example of "Christian virtue and manual labour."[33]

Another important area in the Phoenix hinterland, Riverside, lay about seven miles further north and was home to a community of ex-indentured Indian market gardeners (Fig. 4).[34] In their wake came a variety of evangelists. One such, as Nile Green has shown, was a Bombay holy man, Ghulam Muhammed, who arrived in Durban in 1896. Here he followed a program of madrassa and mosque building along with projects of social welfare and uplift. As part of this work he distributed pamphlets, hagiographies, and miracle narratives that he had brought from Bombay. The lawyer who helped him draw up the title deeds for his land transactions was none other than Gandhi. As Green points out, there are similarities between these men: both had migrated to Bombay from towns in western India and then made their way to Durban, where both relied on the same Muslim and Parsee merchant benefactors.[35]

Another group of evangelists who followed in the wake of the indentured laborers emanated from the Arya Samaj, the Hindu reformist movement established by Dayananda Saraswati in 1875.[36] Driven by a strong evangelical agenda, the Arya Samaj evinced a strong interest in Indians abroad, seeing an opportunity to "save" those lost to superstition and to other religions.[37] Toward this end, Bhai Parmanand visited South Africa in 1905, establishing the Hindu Reform Society in Durban and the Hindu Young Men's Association in Pietermaritzburg. As Goolam Vahed's work shows, he was followed in 1908 by Swami Shankeranand, who, like Parmanand, came from the Punjab, the Arya Samaj heartland.[38] The swami spent just under five years in South Africa concentrating on building a Hindu constituency among "colonial-born" Indians. He sought to "save" this group from Muslim merchants, popular

Islamic influence, especially the Muharram festival, which Hindus and Muslims celebrated, and Gandhi ("a Tolstoyan," in Shankeranand's words, who did not serve Hindus well).[39] Shankeranand established a welter of cultural organizations, promoted Diwali in opposition to Muharram, and endorsed cow protection. While he concentrated much of his energy on Durban and Pietermaritzburg, he also visited and lectured north of Durban in the Phoenix area.[40]

The hinterland around Phoenix constituted a brave new world of evangelical experiment comprising proselytizing Trappists, mid-Western Protestants, Zulu internationalists, Bombay Muslim holy men, and Punjabi Arya Samajists. In this environment different versions of India encountered each other as well as different ideals of Africa. It was an appropriate setting for the IPP.

The Move to Phoenix

Unsurprisingly, for a project of social reform, the founding narrative of Phoenix features the press as a central protagonist. Indeed, the first structure on the settlement was built to house the printing operation. Constructed from donated material with the help of indentured laborers from the Ambulance Corps, the corrugated iron building was an unusual shape for a print shop, being more square than thinly rectangular, the ideal configuration for printing premises to maximize light.[41] This slightly eccentric shape symbolized much of the spirit of the extemporization and experimentation that would characterize the press and the settlement.

With the building partly constructed, the press followed, its journey cast as a venture to the interior. Four wagons, each headed by a span of sixteen oxen, carried the printing machinery and equipment, fording rivers and crossing rugged countryside. The cavalcade arrived at nightfall, camping out like a party of pioneers. In West's account, "Such an encampment, surrounding the half-finished press building, had never been seen there before in living memory."[42]

Central to this party of pioneers were three stalwarts from the IPP: Albert West, Sam Govindswamy, and Chhaganlal Gandhi, who had been working on the Gujarati section of the paper. Along with the indentured workers, they immediately began setting up a print shop in the wilderness, putting up racks, erecting frames, and assembling machinery. Their perseverance paid off, and, despite the disruptive move, the paper continued to appear with just a few editions having to be printed in Durban. On December 24, 1904, the first issue came off the press at Phoenix. This shift from town press to country press was marked by a change in the size of the paper. Instead of eight broadsheet pages, the paper now was made up of sixteen foolscap sheets folded into book form that could, in emergencies, be printed on a treadle machine should there be problems with the other equipment.[43]

Having established the press, the settlers constructed makeshift accommodations for themselves on the hundred-acre property. More settlers soon followed. Gandhi's relatives—Maganlal, Abhaychand, and Anandlal—joined the settlement, as did his friends—Herbert Kitchin and Henry Polak (although from 1905 Polak was summoned back to work in Gandhi's office in Johannesburg). In June 1906, Gandhi's wife, Kasturba, and children arrived. Some settlers also brought their wives and children, while others married and established families. At its height, Phoenix was home to about sixty residents.[44]

The core of this group consisted of "schemers," those who had signed up for the "scheme" by which they were to earn £3 a month, be entitled to two acres of land, and devote themselves to producing *Indian Opinion*. If the press rendered profit (which it never did), this would be shared among the settlers.

With regard to the functioning of the press, Chhaganlal oversaw the Gujarati section and was co-manager of the press with Albert West, while Sam Govindswamy made sure the machinery was in order. By Pyarelal's account, Govindswamy was one of two press employees (the other not named) who retained their original salaries.[45] They were assisted by a dozen volunteers, mostly nonresidents who commuted from Durban. Uma Dhupelia-Mesthrie lists these as Virji Damodar Mehta, Mr. Mannering, Mr. Orchard, Hemchand, Kababhai, Ramnath, Behary, Muthu, Rajcoomar, Harilal Thakar, and a photographer, Brian Gabriel.[46]

After Nazar's death in January 1906, Herbert Kitchin took over briefly as editor. Being a headstrong man, however, he fell out with his colleagues and soon resigned. Polak then assumed editorship but, like Gandhi, was mainly based in Johannesburg, and the paper was hence produced from two centers linked by a flow of correspondence via the overnight Johannesburg-Durban mail train. Through this correspondence and visits to Phoenix, Gandhi involved himself in all levels of press work—giving instruction on how much type to order (*CW* 5: 79) in between recommending time and stress management strategies for the hard-pressed staff (*CW* 5: 81–83, 206–207). With Chhaganlal he debated how to arrive at a charge for advertisements—Chhaganlal wanted to charge on the amount of type used, Gandhi on the space filled (*CW* 5: 145). He also expressed his views on what size the advertisements should be (*CW* 5: 173); how best to split Gujarati words (*CW* 6: 306); and how to design letterheads (*CW* 6: 450).

Printing Utopia

As several historians have demonstrated, the newspaper *Indian Opinion* became important to the success of Gandhi's campaigns.[47] The paper kept people informed, spurred them on, and provided the world with information about what was going on.[48] Equally pivotal was the press itself, which became central to Phoenix and operated as an embodiment of its utopian ideals. Virtually all residents—men, women, and children—were involved in at least some aspect of the printing process. Typesetting was mandatory for all literate members of the settlement, some proving to be more adept than others, with Gandhi describing himself as a dunce.[49] Most men assisted with operating the presses while everyone folded the newspapers, put them in wrappers, and pasted on addresses.[50] Job descriptions were exhaustingly varied, Chhaganlal turning his hand variously to bookkeeping, compositing, translation, editing, gardening, and jungle clearing (*CW* 5: 81-83, 206-207). In addition to being involved in printing and gardening, West did proofreading, office work, reporting, and subediting, and he read the exchange papers as well as involving himself in public projects around Phoenix.[51] In this context, the definition of a job itself could be described in exasperatingly utopian ways. At one point the staff at Phoenix was unhappy with a colleague and wanted to get rid of him. Gandhi insisted that he was well qualified to stay on, since he was celibate and patriotic (*CW* 6: 448).

Among the settlers at Phoenix were several of Gandhi's relatives, as well as his immediate family. The core functions of the press were family run. Alongside Chhaganlal who had wide-ranging responsibilities, Maganlal oversaw composing and did other skilled jobs, while Anandal picked up Gujarati typography. However, this was not a family business in the normal sense. As Prabhudas Gandhi noted in his memoir: "In Gandhi's ashram the place of blood ties was taken by common ideology and a common devotion to duty."[52] Or, as Lloyd and Susanne Rudolph have indicated, Gandhi's ashrams were modern institutions in which entry and exit were by choice rather than lineage.[53] As James Hunt notes, Phoenix stood midway between a village and a joint family.[54] This demotic impulse was strengthened by the press, which functioned as a leveler, with everyone undertaking physical labor, whatever their caste or religious background. In the words of Prabhudas Gandhi, "Germans, English, Africans, Chinese, Parsis, Muslims, Jews and Hindus" labored together on the press.[55]

The press formed a plank in Gandhi's educational philosophy. Part of Phoenix's work involved running an informal boarding school for children both from the settlement and beyond. Prabhudas Gandhi attended this school and left a description of his routine. After an early start, students attended school from 9:00 A.M. to 11:00 A.M. They then spent a half hour digging in the fields. When students complained and asked whether this digging could not happen early in the morning, Gandhi replied: "You must get into the habit of working in the fields in the heat of the sun. Today you are studying here, but if the struggle starts and you have to go to jail, who will then let you rest in the shade?" At 11:30 A.M. the students bathed, had lunch, and worked on their own, while the adults were busy in the press. At 3:00 P.M. the children went to the press and received vocational training or assisted with press work. At 5:00 P.M. the children returned to work in the fields until sunset.[56]

Gandhi saw Phoenix as a training ground for satyagrahis, and for prison. There was hence a rigorous labor regime of which the press formed a part. Working a hand-operated iron press is noisy, and it involves long hours. It is physically demanding, and those who did it on a full-time basis often developed back and kidney problems from the constant strain of pulling the bar of the press.[57] Typesetters also developed aching fingers and felons (painful, pus-producing infections at the ends of their fingers).[58] While the settlers at Phoenix appear to have avoided these ailments, they were nonetheless subject to the labor disciplines of the print shop, which increasingly became part of the project of training for satyagraha.

The press functioned as a fulcrum of social relationships on Phoenix, which was at times known as the IPP settlement. When Albert West married Ada Pywell in July 1908, the wedding was reported in *Indian Opinion* as though the bride was marrying into the printing press (July 4, 1908). The reception took the form of a wedding breakfast, made and served by the men, during which speeches were delivered. The bridegroom's speech discussed Gandhian educational principles and the proposed school at Phoenix, which "would unlock the door of prejudice and unbar the gate of race hatred" (July 4, 1908).

As part of a settlement run on Tolstoyan and Ruskinian principles, the press constituted a kind of religion, the ideal for which settlers lived and worked. Gandhi reminded Chhaganlal that printing was first and foremost about principle (CW 7: 61): "Our capital is not the money we have, but our courage, our faith, our truthfulness and our ability" (CW 6: 252). Press work was never an end in itself, and initially the idea had been that printing would be a part-time activity fit in between agricultural work. As matters turned out, the agricultural ambitions of the settlement never took off, and printing became the main occupation. Religion constituted the other important dimension of community life. Daily routines hence included plowing, printing, and praying. An average day would begin with work in the fields, followed by a long stint in the printing press, and ending with a daily half-hour multifaith service in which "Hindus, Muslims, Parsee and Christians sang hymns and read the various scriptures in different languages," as Albert West described it.[59]

As part of turning printing into a kind of religion, Gandhi argued strenuously for the use of hand-operated technology in which printing could be slowed down and freed from the hasty tempos and alienation of industrial production. Yet in practice this probably happened infrequently, and instead the IPP arrived at a compromise between hand and machine power. An experienced printer, West knew that one could not run heavy machinery on manual labor alone, and so he acquired an oil engine. But to keep the gospel of hand work alive, he invented what was called "the wheel"—a driving wheel with handles on either side that when manually operated powered a long belt, and in turn a machine.

The utopian dimensions of the press were strengthened when in 1910 Gandhi decided to stop all jobbing printing, seeing this as a distraction from the real work of producing the newspaper and its associated pamphlets. In the same year, he began scaling back on advertisements, and then in 1912 he decided to dispense with virtually all advertisements except those promoting socially useful objects, especially books.[60]

The changing letterheads of the press reflect these shifts. An early letterhead from March 1904 when the IPP still operated from Durban presents the press as a commercial operation ("artistic and general printers") undertaking jobbing printing of various kinds (wedding cards, visiting cards, ball programs, etc.) in a range of languages (Fig. 6). By 1907 the letterhead had lost some of its commercial flair (Fig. 7). Thereafter the press's sole function was to print *Indian Opinion* (Fig. 8).[61] Another barometer of these shifts comes from the memoir of Prabhudas Gandhi, who spent part of his childhood at Phoenix. He describes these years of mounting asceticism as a period when there was "no more fun and frolic."[62]

The idea that everyone worked equally on the press was probably something of an illusion. There was a core group of people devoted to the running of the press; two people earned commercial salaries. It also is clear from some accounts that the press, like all concerns in Natal, relied on cheap African labor. Millie Polak reported:

> The printing-press, at this time, had no mechanical means at its disposal, for the oil-engine had broken down, and at first animal power was utilized, two donkeys being used to turn the handle of the machine. But Mr. Gandhi, ever a believer in man doing his own work, soon altered this, and four hefty Zulu girls were procured for a few hours on printing day. These took the work in turns, two at a time, while the other two rested; but every male able-bodied settler, Mr. Gandhi included, took his turn at the handle, and thus the copies of the paper were "ground out."[63]

Gandhi, in his autobiography, notes his preference for a hand-operated press that he regarded as more uplifting: "There came a time when we deliberately gave up the use of the engine and worked with hand-power only. Those were, to my mind, the days of the highest moral uplift for Phoenix."[64]

There is of course an irony when this claim is set against Polak's description indicating that four Zulu women did the hardest physical labor. On the face of it, Phoenix stood beyond the wage economy, since in theory settlers did not receive salaries, only a £3 grant. However, as we have seen, two members of the press did receive commercially linked salaries, and so the press at least was not entirely extricated from the relations of the market. One does not know what the Zulu women were paid, if anything. Their absence from Gandhi's account points to the extent to which his press was a South African one that relied, if only in small part, on cheap African labor.

The final utopian dimension of the press pertained to copyright practice. While Gandhi objected to plagiarism, he initially regarded copyright law as a form of private property that prevented the free circulation of ideas.[65] Two of the pamphlets produced by the IPP (*Hind Swaraj* and Tolstoy's *Letter to a Hindoo*) explicitly indicated "No Rights Reserved." In following this policy, Gandhi was in effect seeking a way of operating not only beyond the constraints of the market but of the state as well. In this respect he outstripped many of the evangelical presses in the Durban hinterland: records show that Christian mission organizations generally observed copyright legislation studiously.

Avowedly cosmopolitan in its personnel, methods of working, textual products, and envisaged audiences, the IPP played a central role in the construction of this utopian vision. Commenting on the wide range of people who visited Phoenix, Prabhudas Gandhi noted: "Our jungle school had the atmosphere of an international university."[66] His observation points both to the brave experiment of Phoenix as well as its limits marked here by the jungle, and its implied African inhabitants. Like the Zulu women erased from Gandhi's account of the press, Africans were not numbered among the fraternity of Gandhi's cosmopolitanism.

Yet in the hinterland of Phoenix such limits were not unusual. Here, "race-making" projects like Gandhi's and Dube's shaped ideas of Africa and India wrought in relation, and, in opposition, to each other. While easy to judge in hindsight, these undertakings form an important strand in shaping ideas about "race" that, as Tony Ballantyne, Antoinette Burton, and Jonathon Glassman have argued, were not the sole prerogative of European policy makers, triaging subjects by race and civilization according to the administrative logics of the imperium.[67] The Indian Ocean world contained a rich archive of precolonial ideas of "race," culture, and civilization that meshed with, and challenged, colonial categories, creating a hierarchical world in which utopian ideals of "race" could be both anticolonial vis-à-vis white imperial interests and colonial in relation to "also-colonized others."[68]

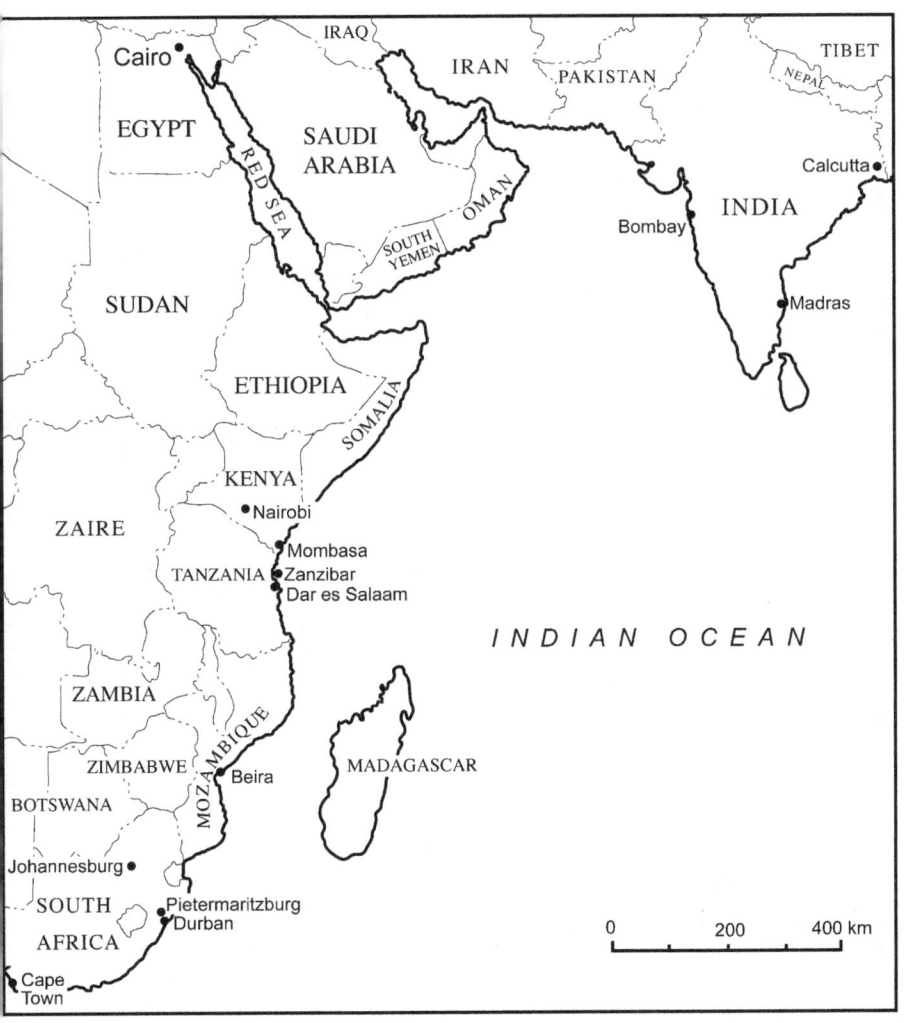

Figure 1 Map of the Indian Ocean, showing major cities mentioned in
the text. Reproduced with permission from the University of the
Witwatersrand.

Figure 2   Map: Printing establishments, Durban, 1903. Reproduced with permission from the University of the Witwatersrand. The black dots represent white-run firms. The white dot in Grey Street shows the position of the International Printing Press. The second white dot represents the *Indian Opinion* office.

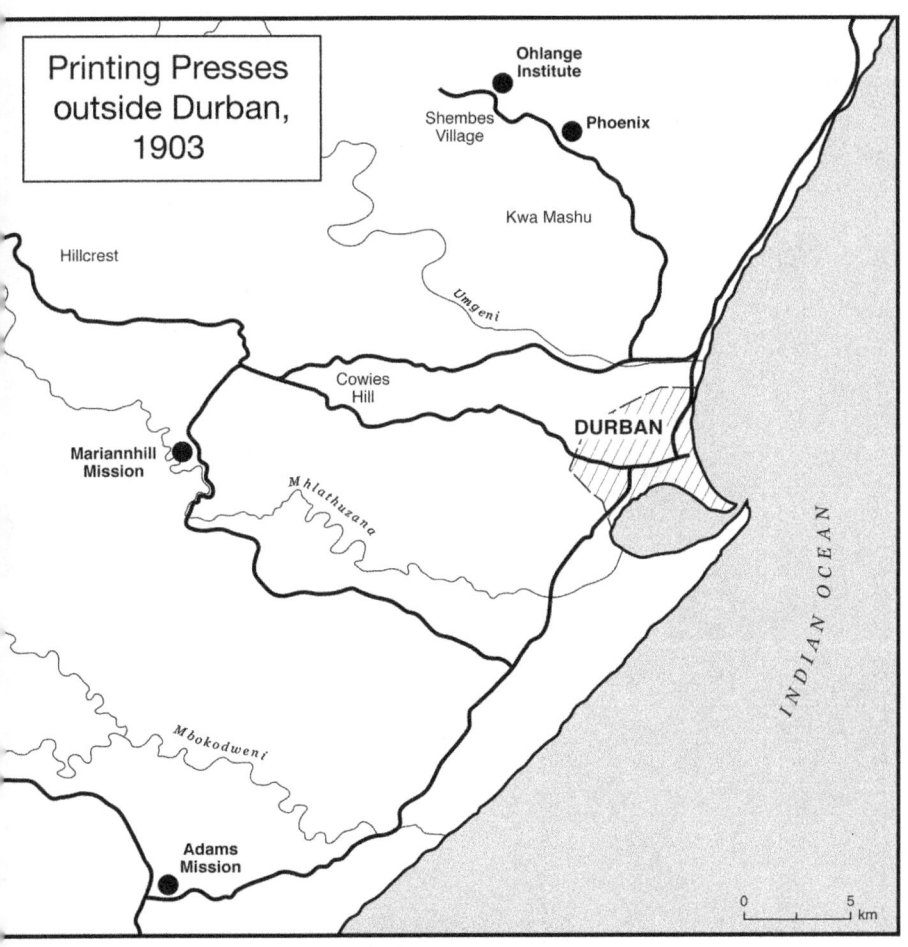

Figure 3   Map: Printing presses outside Durban, 1903. Reproduced with permission from the University of the Witwatersrand.

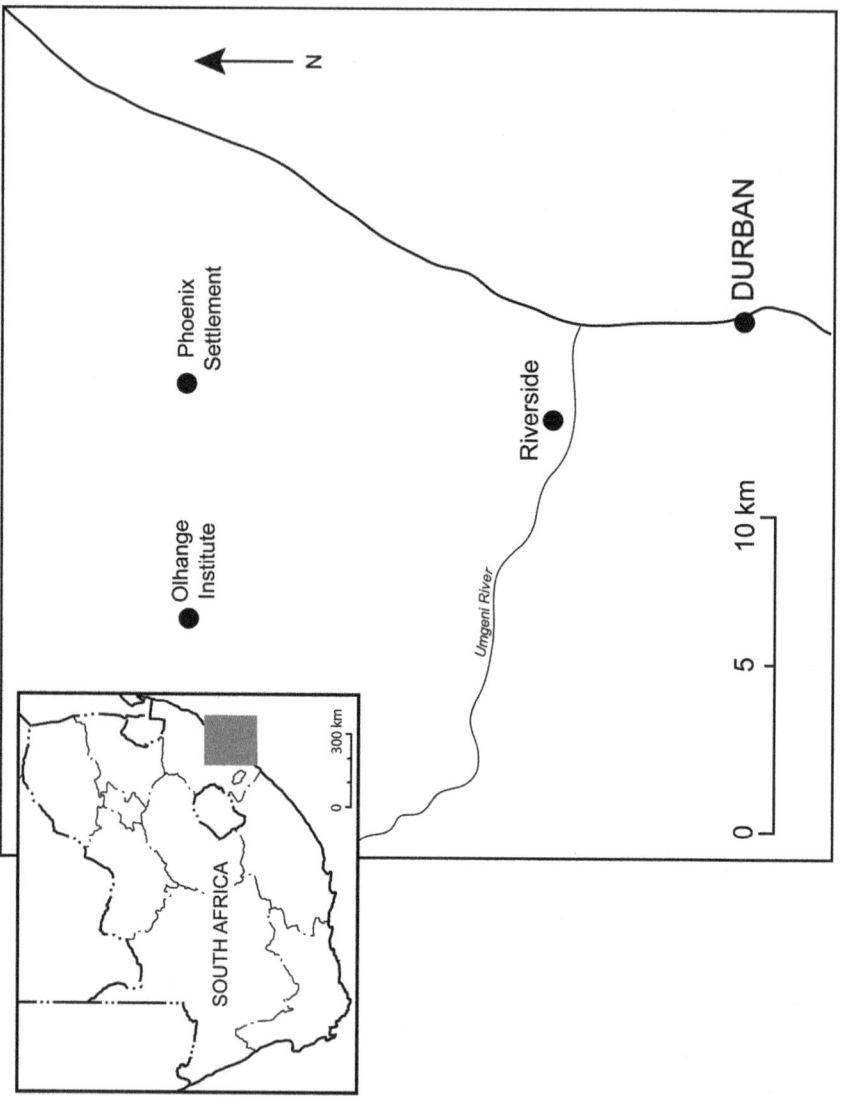

Figure 4   Map of Phoenix hinterland. Reproduced with permission from the University of the

Figure 5   Photograph of lynch mob gathered on Durban dockside awaiting Gandhi's arrival, 1897. Reproduced with permission of the Local History Museums' Collection, Durban.

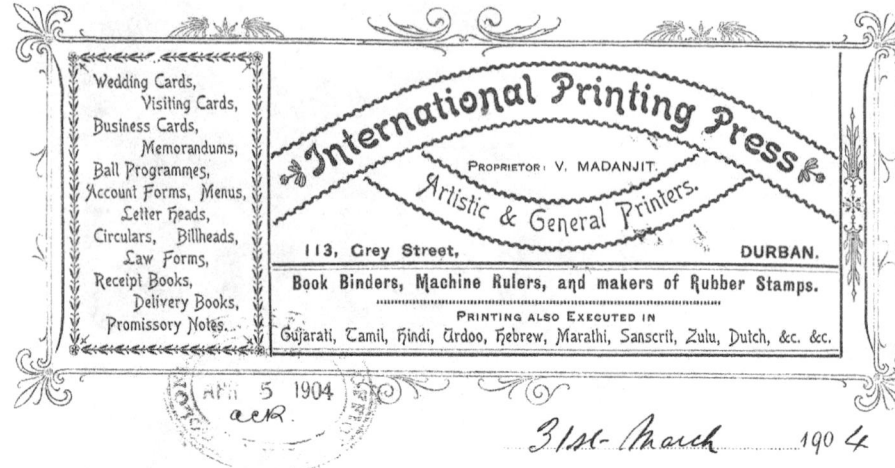

Figure 6   Letterhead for the International Printing Press, 1904 (NAD CSO 1758 1904/2954). With acknowledgments to the Pietermaritzburg Archives Repository.

Figure 7   Letterhead for the International Printing Press and *Indian Opinion*, 1907 (NAD CSO 1848 8564/1907). With acknowledgments to the Pietermaritzburg Archives Repository.

Figure 8   Letterhead for the International Printing Press and *Indian Opinion*, 1913 (NAD II 1/180 I 1058/1911). With acknowledgments to the Pietermaritzburg Archives Repository.

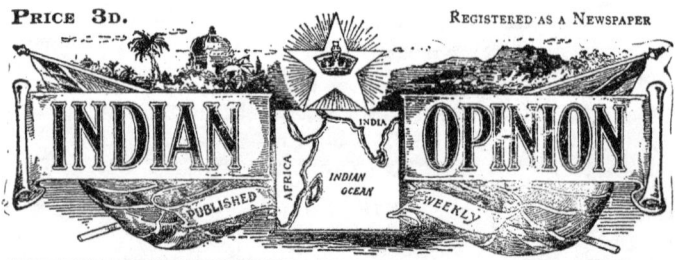
Figure 9   Masthead of *Indian Opinion*, December 5, 1908.

3

Indian Opinion

Texts in Transit

Late in 1909 Gandhi issued a letter of instruction to the staff at Phoenix. The newspaper was becoming a strain, and its size was to be reduced. "All matter should be severely condensed. Energy should be devoted to the art of condensation.... Original papers from which condensation is made should, if possible, be kept pasted in book form" (*CW* 10: 195).

A year later, in 1910, the paper again downscaled but reassured its readers: "Though the size has been reduced, we hope we shall be able by means of condensation to give the same amount of information" (*Indian Opinion,* January 1, 1910). Three years later, as the satyagraha campaign took its toll, the paper once again curtailed the number of pages. As it was explained in the Gujarati section:

> It is our intention to continue providing the same [reading] matter [as before], but in as short a form as possible. By so doing we will be able to fit in more material within the same space or even less. Beginning this time, we have reduced the number of Gujarati and English pages, but we wish to provide more

information, though not more words within these pages. It is our hope to reduce the work of the compositor while increasing that of the writer. (*CW* II: 249)

In making a virtue of necessity, this proposed textual dispensation is characteristically Gandhian: loss is converted to gain, and less turns out to be more. The normal order of things is reversed. The compositor's work becomes light, and the writer's work becomes dense and demanding. The weight of the type seems to sink into the printed words themselves, making them heavy with concentrated meaning.

The characteristics of journalism—telegraphic brevity, rapid summary, hurried reading, instantaneous obsolescence—become their opposite: summary acquires a gravity that pulls the reader's attention down into the text where reading has to be deliberate rather than thoughtless and hasty. Condensation becomes an art form that produces a thoughtful or an "ideaful" text that in turn requires a reader who is thoughtful in both senses of the word: exercising careful deliberation and extending sympathetic regard to the text. The art of condensation requires consideration—from the writer, the reader, and from us, a contemporary audience accustomed to seeing summary, abridgment, or précis as trivial intellectual procedures rather than as a form demanding particular orders of skill and craft.

In his autobiography Gandhi devotes a chapter to *Indian Opinion*, which reiterates this theme of the newspaper concentrating language. "Week after week," he says, "I poured my soul in[to] its columns." Every word entailed "thought or deliberation." Any superfluity—"a word of conscious exaggeration, or anything merely to please"—was trimmed. The depth of condensation is considerable:

the passion of the soul itself is pared down to spare prose. This labor, however, has ethical benefits, since writing in this way proved to be "a training ground in self-restraint."[1]

Gandhi has long been celebrated for his plain prose style, a product of such textual self-restraint. This mode of composition with "not one word more than necessary" constituted a form of spiritual discipline.[2]

> The reader can have no idea of the restraint I have to exercise
> from week to week in the choice of topics on my vocabulary. It
> is a training for me. It enables me to peep into myself and make
> discoveries of my weakness. Often my vanity dictates a smart
> expression or my anger a harsh adjective. It is a terrible ordeal but
> a fine exercise to remove these weeds.[3]

Indian Opinion provided one matrix in which this style took shape. Like most of its contemporaries in colonial settings, the journal was largely generated through cuttings and summaries gleaned from exchange papers from far and wide. Transnational summary hence constituted much of the paper's substance. The staff made ingenious use of this format to create a niche and identity for the paper. Gandhi's particular contribution was to extend these methods into his ethical writings for the paper. Through summary and abridgment of texts from across the world, he instituted new kinds of ethical "exchanges," helping readers see themselves as part of a vast reading commonwealth while making the ethical extracts and the "news" clippings among which they were placed equivalent to each other. This juxtaposition redefined both genres, making ethical discourse "news" and slowing down news reports to the pace of philosophy.

Indian Opinion: Historiographies

Indian Opinion is one of the great intellectual archives of the world. Produced in a context of multiple diasporic intersections in southern Africa, the paper's pages are woven from a variety of global intellectual filaments, drawn from larger trajectories of migration. These intellectual maps include the imperial and subimperial triangle of Africa-India-Britain; the dispersal of indentured Indian workers (Mauritius, Fiji, Caribbean, Burma, Ceylon, Malaya); the flashpoints where voluntary Indian migration encountered the global color line (British Columbia, the United States, Australia, New Zealand); the sacred geographies of Islam and Hinduism, followed by Zoroastrianism, Jainism, and Christianity; circuits of black internationalism (Universal Races Congress, Booker T. Washington, W.E.B. Du Bois); and configurations within and between other European empires (Portuguese, French, German).

In any given issue or, indeed, on almost any given page, these different strands are plaited and crosshatched, their juxtaposition provoking new intellectual possibilities. The English section of the October 21, 1911, issue includes articles on:

- a celebration in Mozambique of the first anniversary of the Portuguese Republic;
- a meeting of the Pietermaritzburg Hindu Young Men's Association;
- "The Position of British Indians in the Dominion";
- the upcoming Transvaal Women's Indian Association Bazaar;
- "The Jewish Home" (a report on the Jewish family as foundational, from *The Sentinel*);

- "The Coloured Man in Art and Letters" (Pushkin, Dumas, Dunbar, Coleridge-Taylor, Will Marion Cook, Du Bois, Washington, Blyden) (by Duse Mahomed, from *T.P.'s Magazine*);
- the conference of the Native (African) Women's Christian Union (taken from the *Transvaal Leader*);
- "Mahomedan Wives" (a report on a case in the Transvaal that ruled that Indians could have only one wife, taken from *The Comrade*, Calcutta); and
- Rev. E. J. Moqoboli who received an honorary doctorate from the Council of Education in the United States (taken from *Imvo Zabantsundu*).

These pieces are interspersed between reports on legislation, violence, and evictions directed against Indian communities in various parts of South Africa as well as ongoing attempts to keep satyagraha alive.

In the world of periodical journalism, where such global bits and pieces were standard fare, the diversity of this lineup was not entirely unusual. Yet there were probably few papers in which one would encounter this particular intersection of empires, races, and religions that requires one to think about the cantata *Hiawatha* (by composer Samuel Coleridge-Taylor) and Dumas's *The Three Musketeers* in between Muslim marriage in the Transvaal, the African Christian elite in South Africa, and the details of the arrangements for the upcoming bazaar in Johannesburg.

This mode of compiling transnational portfolios is also evident in the book pages that became an increasingly important feature of *Indian Opinion*. Initially limited to advertisements for books from India (mainly those produced by *Indian Review*, January 7, 1904, and self-help books, November 19, 1904), these columns eventually occupied most of a page, filling some of the space vacated by advertisements on which Gandhi began to scale back from 1912. Alongside a weekly almanac setting out the Christian, Hindu, Muslim, and Parsi calendars, the book pages, in a mixture of English and Gujarati, offered several dozen volumes for sale via the International Printing Press (IPP). In addition to the full range of English and Gujarati pamphlets produced by the IPP (the subject of the next chapter), the pages offered several Gujarati titles imported from India. Its English offerings included Gandhi's favorites, namely, Tolstoy *(Tales and Parables, War and Peace)*, Ruskin *(Sesame and Lilies, Time and Tide, Unto this Last)*, and Emerson *(English Traits, Representative Men, Society and Solitude)*, as well as Mazzini *(Duties of Man)*, and Burke *(Reflections on the French Revolution)*.[4]

A second category consists of works on theological, religious, and ethical themes: *Rubaiyat* of Omar Khayyam, *Gita, Ramayana, Confessions of Al Ghazzali, Persian Mystics,* Sa'di's *Gulistan or Flower Garden,* Meredith Townsend's *Mahomed "The Great Arabian,"* Washington Irving's, *The Life of Mahomet,* and an English version of the Koran. As works on Gandhi began appearing (Joseph Doke's biography, *M. K. Gandhi: An Indian Patriot in South Africa,* and P. J. Mehta, *M. K. Gandhi and the South African Problem*), these were listed, as were portraits and images taken from the newspaper itself and offered for sale as separate items.

While my account has sorted these books into categories, in the book pages themselves they appear as long lists, and one can hence move from the South African writer Olive Schreiner to the anarchist Kropotkin, to the Victorian self-help guru Samuel Smiles, to a work on the Japanese social reformer Ninomiya Sontoku, *The Peasant Sage of Japan.*

As scholars of periodicals have indicated, formulating ways of analyzing this abundance is never easy.[5] In the case of *Indian Opinion* these problems are more acute, since there has been little interest in analyzing this material from a literary perspective. Scholars have used the paper as a source of historical data, and there have also been important discussions of the paper as an instrument of satyagraha, most notably Uma Dhupelia-Mesthrie's history of the paper and her outstanding biography of Manilal, Gandhi's son, who took charge of the press and *Indian Opinion* after Gandhi's departure and kept it going, editing the paper until his death in 1956.[6]

In the absence of any attempts to analyze the paper from a literary perspective, this chapter begins at the beginning, examining the bread-and-butter form that made up the bulk of the paper, namely, the summarized newspaper clippings. These clippings are one of the first things that strikes one about the paper, and at times, especially when satyagraha imprisonments had taken their toll, made up half to three-quarters of the content of the paper (a fairly good ratio, it must be said in the world of colonial periodicals, where sometimes an entire publication comprised only clippings).

Mails and Exchanges

In the fabric of the newspaper itself, this world of exchanges is taken for granted, something known to both editor and reader. Mentions of "our South African exchanges" or "our Indian exchanges" are casually sprinkled into stories (February 11, 1905; January 20, 1906; April 23, 1910; February 25, 1911; April 20, 1912; March

18, 1914). Likewise, the "Indian mail" or "English mail" (another term for the exchanges) is always "to hand," as the articles phrase it, implying that both writer and reader are familiar with the practice and its genres (May 20, 1905; October 27, 1906; January 22, 1910; February 18, 1911; April 8, 1911).

The shorthand quality of these references implies knowledge of a broader world of circulation that is habitual and taken for granted. The term "Indian mail" or "English mail," for example, compresses a number of meanings: the mail boat itself; the post and newspapers that it carried; and possibly the post office that might have forwarded these (say from Cape Town) and the railway that carried them, thereby establishing a common link between post and news as continuous institutions.[7] Reminders of these multiple meanings appeared elsewhere in the paper. Each week a small piece entitled "Indian Mail" detailed the routes and departure dates of the Indian Ocean mail boats: Bombay direct; Bombay via the East Coast (of Africa); Colombo (with transfers to Bombay), Calcutta, Madras, and so on.

In the early years of the paper, this snippet was generally sandwiched between the editorial below and "Subscriptions" above. The editorial appeared under a miniature masthead and the date of the edition; "Subscriptions" listed the different rates for subscribers in southern Africa, India, and England. The stacking of these three items (Subscriptions, Indian Mail, and Editorial) implies a familiarity with transnational modes of reading that assume circulation as a normal framework. *Indian Opinion* hence emerges as a paper made in circulation, carried *by* Indian and English mail to subscribers elsewhere, and composed *from* Indian (and other) mail from afar. It is as though "Indian Opinion" (in both senses of the phrase: the newspaper and Indian public opinion in South Africa) emerges from the gap between these two sections (Subscriptions and Indian Mail), creating something new out of incoming and outgoing matter.

Exchange-Made Worlds

Indian Opinion made imaginative use of this exchange system, dramatizing various points and nodes of circulation to create stories for the paper.

One such node constituted the arrival of documents from elsewhere, an event that constituted something of an occasion. When the English mail boat dropped anchor in Durban harbor, for example, a gun would be fired to signal that the post had arrived. The columns of *Indian Opinion* specialized in spotting, using, and creating such "occasional texts" (both written for and creating an occasion). When important periodicals like *Indian Review* (which served as a model for *Indian Opinion*) reached Durban from Madras, the journal's arrival constituted a story in its own right, and its content was summarized for readers. When *Indian Opinion* staff published pieces in prominent journals, as Polak did in *Empire Review*, the event became a report in the paper. Indeed, one minor calling card of the paper comprised *Indian Opinion* quoting itself being quoted by others, an external validation of the paper's identity or what its columns referred to as "ourselves" (October 10, 1904; February 4, 1905; July 15, 1905; December 30, 1905). *Indian Opinion* itself was something of a weekly event, with subscribers in Johannesburg, at least, according to Gandhi, eagerly awaiting their copies and disappointed if there were transport delays.[8]

The very idea of writing (at least in prose) becomes associated with this textual flow of the exchange papers. The figures of the writer and the intellectual are frequently invoked in the paper, but they seldom appear under their own byline, or in the guise of a "genius" expressing some irreducible subjectivity. Instead, writers enter the paper via the clipping: G. K. Chesterton discusses "What Is Indian Nationalism?" via the *Illustrated London News* (November 13, 1909); Booker T. Washington comes via a piece by Romain Rolland in *East and West* (September 10, 1903); and Tagore speaks through a medley of papers, including *The Observer, Christian Commonwealth,* and *Modern Review* (July 12, 1913). Even a local writer like Olive Schreiner has her ideas on *Closer Union,* made "viable" through the *Transvaal Leader* (January 2, 1909). Authors and intellectuals also appear as headlines "Keshub Chunder Sen" (February 4, 1905) and "Babu Surendra Nath Banerjee" (June 2, 1906). The writer becomes an event rather than a byline.

Yet within the world of exchange circulation, genres move in different ways and at different imaginative speeds. Unlike prose, poetry appears able to free itself from the textual relay of the exchanges, and poems arrive unmediated by other publications and are placed above the signature of the author in the English section or with a "signature line" in the Gujarati section.[9] Judged against the slow exchange-paper chain, poetry appeared more immediate, rising directly off the page. Poetry melted distance, and poems played an important part in establishing networks of solidarity between South Africa and India. The poem "The Cry of the Transvaal," published in South Africa and India (*Modern Review,* June 1910, and *Indian Opinion,* August 6, 1910), creates the impression that the lament of oppressed South African Indians resonates so acutely in the ears of Indian sympathizers on the mainland that one might think they occupied the same house.

Prose writers, however, have no such freedom, existing in and arising from the world of periodicals. Even a byline-worthy literary "big man" like Tolstoy, a frequent inhabitant of *Indian Opinion,* appears less a unique genius whose thoughts can only be admired than an ethical writer whose ideas can be extracted, emulated, and followed. The disclaimer "No Rights Reserved" that appeared at the beginning of Tolstoy's *Letter to a Hindoo* underlines this point by naming the author not as a legal persona but as an entity created via the free circulation and application of his or her texts.

In some rare instances, *Indian Opinion* did use bylines, but, again, this was tied to occasions of circulation in which the paper sought to dramatize its networks. In 1905 Bhai Parmanand, an Arya Samaj missionary, visited Natal and established the Hindu Young Men's Association. As a representative of a transnational Hindu reformist organization with a small but elite following in Natal, he merited his own byline—"Professor Parmanand" (December 2, 1905). In the same year, J. L. P. Erasmus, a Boer commandant held as a prisoner of war in India during the Anglo-Boer War, wrote for the journal. He received a conspicuous byline and a profile from the paper wishing to show its links to "well-intentioned" whites (January 28, 1905).

With regard to its own staff, the paper's stated policy was that "we write impersonally" (December 24, 1904). Yet, like many other periodicals, forms of authorship, such as fictitious correspondents and pseudonymous and semipseudonymous writing,[10] proliferated, in part to make the journal seem larger than it was. In a world where circulation conferred visibility, these forms of authorship were called into being by travel itself, which furnished one opportunity for staff to emerge from anonymity. On his trip to

India, Polak wrote an account of his sea voyage under (or, rather, above) his own name; West's travelogue of his trip to England appeared with his initials attached.[11]

Gandhi himself created his own textual persona by circulating in and out of the journal through letters, signed articles, cartoons, and reports on him from other papers. With regard to letters, Gandhi's correspondence with state officials appeared verbatim in the columns of the paper.[12] In other instances Gandhi wrote letters *to* the editor on issues of grave importance, such as the perfidy of his nemesis, Jan Smuts, or the "turncoat" Ram Sunder Pandit, an early hero of the campaign, who did one jail term and then abandoned the movement (*CW* 7: 113, 254).

In the signed article, "M. K. Gandhi" or "Mohandas Karamchand Gandhi" makes important announcements regarding satyagraha, or attempts to dispel misunderstanding, especially around the compromise on fingerprinting in January 1908 when followers were asked to submit voluntarily to the requirements of a law that previously they had been asked to resist. These two signatures are distributed in slightly different ways across the English and Gujarati sections. In the English section, the calling card is always "M. K. Gandhi," the tone polite, the form a letter. While such genres occur in the Gujarati section, there is a new and distinctive voice, "Mohandas Karamchand Gandhi," who speaks both through letter and the signed article in a voice that is outspoken, sometimes hectoring, emotional, passionate, truth-telling, and personal. This voice explains painful decisions like permitting his sons to go to prison (*CW* 9: 42–43); defends against unjust attacks from within the community—from Pathans, from Muslim merchants accusing him of ruining them (*CW* 9: 258–260), and from those critical of the compromise over fingerprinting (*CW* 8: 350–352). This impassioned voice calls for sacrifice and support and confronts Hindu/Muslim disunity (*CW* 8: 160–162). As these Gujarati articles proceed, this persona increasingly acquires the identity of satyagrahi, as the ever-more-elaborate salutations at the end of letters indicate: "I remain, satyagrahi"; "Servant of the Community and Satyagrahi"; "India's servant, Satyagrahi; I am India's bond slave; I am, as ever, the community's indentured labourer" (*CW* 9: 43; 14: 26, 205).

Beyond these self-created textual voices, Gandhi entered the newspaper via endless *Indian Opinion*-generated reports on his actions and speeches, as well as clippings from elsewhere. As a prominent political player, Gandhi made his way into numerous cartoons in South African newspapers, and these were reproduced in *Indian Opinion* (and explained by Gandhi to his Gujarati readers) (*CW* 8: 36, 135, 142; 9: 24). Through these he emerges as an ever-more-prominent character, coming in and out of the paper on a weekly basis.

Times of Circulation

Because of the necessary time lags involved, journalism by exchanges lacked the insistent tempo of telegraph-driven and dateline-dominated reporting. Instead, it created slower and more leisured genres. Some stories coming firsthand (or secondhand, via an exchange) from Reuter are datelined, but these are few and far between.[13] Local news reports generally (but not always) have dates embedded in the story, but virtually all of the exchange material is undated.

The paper exploited this feature to create a sense of more leisured tempos of reading. As we have seen earlier, *Indian Opinion* included essays, views, and opinions from writers and intellectuals across the world, but generally via some other publication. Not only did such forms have a longer life than news, but evidence of their prior circulation added to their value, ability to speak to different audiences, and durability, making subscribers feel like part of an enduring commonwealth of readers.

This delayed flow of intelligence via the exchanges also informed the genres produced by staff members on the paper. Writers on the staff, especially Gandhi and Polak, specialized in genres like the letter and the note to convey news in a less temporally driven form. Columns like "Our London Letter," "Durban Notes," and "Johannesburg Letter" portrayed these cities as best captured in forms that metaphorically amble along the usual pathways of circulation rather than hurtling along via telegram and telegraph. Another slow mode was the silhouette, which implied contemplating an outline. Polak used the form to profile a series of types in the white community in Johannesburg. His column highlighted the range of contradictory views within one group, unlike pieces in the white press (especially "Man in the Moon" in the *Natal Witness*) that portrayed the Indian community through a leveling stereotype.[14]

"News" in *Indian Opinion* in fact seems more like a stream of opinion, idea, and belief, culled from other papers, than a portfolio of occurrence. In this environment the definition of an "event" itself comes under revision, dissolving into a kaleidoscope of cuttings. A report under the heading "The Bengal Partition and the Swadeshi Movement" consisted of a series of cuttings from *Calcutta Weekly Notes, Pioneer, Amrita Bazar Patrika, Indian Nation, Indian Echo, Indu Prakash,* and *Behar Times,* as well as the *Manchester Guardian* and a cartoon from the *Saturday Westminster Gazette* on November 4, 1905, and "Happenings in India" on April 16, 1910. Such pages create a vivid impression of an event as a configuration of newspapers rather than as a datelined and distinct episode.

In some cases the normal practice of attributing a cutting to a specific newspaper is omitted in favor of assembling a series of snippets under the blanket credit of "the exchanges." A report "Boycott Pars" (paragraphs) compiles five unattributed vignettes taken "from our Indian exchanges" on Bengalis refusing to buy foreign-made goods and coercing others to do likewise. One paragraph tells of a student purchasing some porcelain cups and saucers from a shop on College Street (Calcutta). Discovered by his fellow students, he is compelled to return them, and to donate the refund to a swadeshi cause. In another snippet a Sudra man, carrying his newly purchased pair of Dawson shoes under his arm, is spotted on the *maidan* by some Brahmins who insist he return the shoes. The shop owner refuses to take them back, so the Brahmins refund the man and then "partition" the shoes, tearing them up and casting the fragments "over the heads of [the] grinning crowd of spectators" (October 7, 1905).

Reading such undated and unattributed sequences creates multiple senses of time. The reader both feels himself or herself to be in the midst of these incidents as they are occurring. Yet the reader would also know that this news has taken between a week and a month to reach Durban, depending on the port of embarkation and the route. From this perspective the report acts as a time capsule, releasing its presentness wherever it is published, with the characters on College Street and the *maidan* reenacting their exemplary tableau time and time again. The space-time continuum between South Africa and India could take intricate forms.

South African–Indian Space-Time

While cuttings in the paper were drawn from across the world, the majority were culled from South African and Indian publications. South African titles like the *Vryheid Herald, Bloemfontein Post, Imvo Zabantsundu, East Rand Express, Rand Daily Mail, Pretoria News, Ilanga lase Natal, Ladysmith Gazette, Natal Mercury,* and *Transvaal Leader* jostle for space with *Jame Jamshed, Indu Prakash, Hindu, The Gujarati, Amrika Patrika Bazar, Madras Times, Behar Times, Mahratta, Indian Spectator, Modern Review,* and *Indian Review*. These titles cumulatively build up a sense of South Africa and India as interrelated textual "continents" between which there is a space-time continuum.

Read today, this interrelationship can be confusing, and one is often momentarily disorientated as time and space between South Africa and India combine in unexpected ways. A headline "Loyalty of Native Chiefs" refers not to local African rulers as I first assumed but to the princely states (July 30, 1903). An article with the headline "The National Congress and Indians in South Africa" indicated that a meeting was to be held at "Tata Mansions, Waudby Road" (December 13, 1903). Are we in Durban or Bombay?

Likewise, in terms of time, readers have to be able to imagine interlocking time schemes. One report on November 19, 1903, reads: "By the time this issue of 'Indian Opinion' reaches India, preparations for the meeting of the national assembly [in India] will have very far advanced." Another example emerges from a report in the September 26, 1910, edition: "Mr GA Natesan writes in the *Indian Review* [a Madras publication]: A cable from South Africa brings the news that the British Indians in the Transvaal are taking a vow of passive resistance as a protest against the recent Asiatic Amendment Bill." While the cutting, and others like it, validates local struggles by demonstrating international interest in the event, it also makes it seem as if the event were happening in India and South Africa at the same time. Such simultaneous reporting on satyagraha was common, with locally generated accounts of developments placed alongside clippings on the subject taken from Indian newspapers.

The short piece "Retribution" by Gandhi demonstrates the dense comparative South Africa-India weave that clippings could create (April 23, 1910). Developing a theme close to his heart, the article argues that the oppression of British Indians in South Africa has befallen them as a retribution for caste injustice in India. The piece refers the reader to a clipping from a Calcutta paper that appears elsewhere in the paper. This discusses a meeting in Madras of the Depressed Classes Mission that likened the practices of caste oppression to "a Transvaal within India" itself. Gandhi then goes on to invoke "the Indian exchanges" that bring reports that the upper castes have been complaining about the Gaekwar of Baroda, who had been admitting Pariah boys to the public schools. "We who resent the pariah treatment in South Africa will have to wash our hands clean of this treatment of our own kith and kin in India whom we impertinently describe as 'outcasts.'"

In this piece, links between Transvaal/South Africa and India reach baroque dimensions. Not only are the two connected, but they contain miniature versions of each other within themselves— there is a Transvaal in India, just as one of India's "essences," caste oppression, plays itself out allegorically through racist oppression in South Africa. The intricate interweaving is enabled by a habit of reading and writing shaped by the exchange clipping.

Trained by the conventions of exchange journalism to read transnationally, *Indian Opinion* readers probably navigated such transitions effortlessly. Since the ideal reader was an imperial citizen, he or she would have been accustomed to reading within the time-space continuum of empire, a skill in part built up through regularly negotiating the textual weave of exchange papers. Yet this imperial reading was not simply a matter of charting the flow between metropolis and periphery; more importantly, it was about becoming accustomed to lateral linkages among colonies and non-metropolitan centers.

This particular model of reading informs the masthead, which the paper adopted from the middle of 1908 (Fig. 9). Dominating the image are two scroll-like banners bearing the title. Above these we see the outlines of a horizon, the left side being India, with palm trees and what looks like a colonial building in British-Mughal idiom, and on the right we have Africa represented by an empty but rugged landscape looking a bit like Table Mountain in Cape Town. The left side is clearly possessed of "civilization," the right side devoid of it. Below them billow two crossed Union Jacks united by a map of the Indian Ocean and an imperial crown. These two regions ("civilized" India and empty Africa) and the vast ocean between them are linked horizontally by the periodical and vertically by the imperial crown.

Within this bigger map of India/South Africa/empire, a series of submaps is etched out through weekly reiteration. Weekly advertisements for shipping lines carve the social equivalent of a neural network through their repetition of particular routes like the East African Service—Durban, Delagoa Bay, Beira, Chinde, Mozambique, Dar-es-Salaam, Zanzibar, Tanga, and Mombasa. This circuit draws into its ambit Gujarati merchant networks (old and new), the string of British Indian communities up the East African Coast, and possibly Goan movement between Portuguese East Africa and destina-

tions farther north (February 5, 1910). The Mauritius route—
Durban, Delagoa Bay, Beira, Port Louis, and back again—likewise
repeats on paper the links between these two plantation economies
(Natal and Mauritius), cemented by the flows of skilled personnel,
technology, some indentured labor, and passenger (voluntary
migrant) traffic (June 15, 1912).

Merchant companies likewise advertised on a weekly basis, announcing their reach both within the Indian Ocean and beyond.
The Sindhi company, Pohoomull Brothers in Cape Town, formed
part of a sprawling export empire specializing in "all fashionable
goods, silks of every variety," with branches, as the advertisement
reminded readers, "in India, Egypt, Gibraltar, Malta, China, Japan,
Manilla, Zanzibar, Beira, Salisbury, &c., &c." (March 25, 1911).[15]
Another map was the "South African Directory of Indian Merchants," which appeared weekly, listing companies across South
Africa, Basotholand, Rhodesia, and Mozambique (July 20, 1907).
These trade networks parallel the exchange-clipping geographies
within the newspaper itself, creating a layered world of migratory
and intellectual passages and pathways.

Movement along these routes brings political geographies into
being and confers visibility on political figures. When G. K. Gokhale,
the Congress notable, and Gandhi's mentor, visited South Africa in
1912, the newspaper hyped up his visit by reporting his elaborate
farewell in Bombay and then his reception in Mombasa, Zanzibar,
and Beira, as well as in South Africa. When Gokhale returned, his
progress from port to port and the receptions that he again received were covered.[16] In such travelogues, each stop in a port city
confers value and visibility and promotes the idea of the political
celebrity. It is as if the person is invented through travel, becoming
more "real" and visible with each successive newspaper report.

Through reiterating such routes, the paper habituated readers to geographies, old and new. At times they were required to maneuver within these. "Our correspondent in Trinidad (says *India*) has sent us a number of copies of the local island newspaper the *Mirror* which contains a report [of an overseer on a sugar plantation who shot an Indian labourer]" (May 20, 1911). Readers not only have to pause momentarily to map these connections that arise from vectors of indentured labor migration but also have to reverse its historical order (Trinidad-Calcutta, rather than the other way around). Other pages assemble a series of dots, implying fascinating but as yet unrealized intellectual circuits, and require an almost playful speculation, for example, the *Vryheid Herald, Johannesburg Star, Natal Mercury, Indian Mirror, South Indian Mail, Rangoon Times, South African Jewish Chronicle,* and *Cape Argus* (March 25, 1911).

The ubiquitous use of cuttings created a comparative climate within the pages of the paper, establishing a cumulative archive of important political data within which readers had to draw connections and comparisons. Reports from the flashpoints of racist exclusion like British Columbia, Australia, and New Zealand bring into view the workings of the global color line whose policies and legal manifestations the paper covered in detail.[17] As a subgenre the paper also reported on officials exercising their racist authority to its punitive limits: a Johannesburg zookeeper excludes Indian visitors (July 29, 1907); a captain in the Straits Settlement relegates Chinese first-class passengers to the lower decks (September 17, 1910); and in Australia shipwrecked men of color are refused entry (February 4, 1904). These stories provide an important Gandhian addition to the broader discussion of the global color line, making the concept less an abstract idea than a set of embodied practices residing in individuals who have choices over their decisions.

Ethical Exchanges

In a column "Items of Interest" (April 22, 1911) one meets the usual medley of exchange clippings from the *Agricultural Journal, Natal Provincial Gazette, East Rand Express, India,* and *Hindustan.* Tucked underneath these snippets is a quote from Thoreau, his name appended at the end of it. The quote may in fact have been dropped in to fill a blank space, but its inclusion in a compilation of clippings illustrates well the ways in which the newspaper's ethical emphasis intersected, both in form and content, with the broader practices of the paper. These interventions were largely driven by Gandhi and formed part of his view of the paper as a kind of ethical anthology. Toward this end he excerpted a range of authors like Socrates, Ruskin, Thoreau, Emerson, and Tolstoy. In their physical form these excerpts resemble the columns composed from exchange material: both are a compilation of snippets.

This use of summary and excerpt forms part of Gandhi's project of slowing down reading, of turning the format of the periodical and the newspaper (designed for speed) against itself, and of using the art of condensation to pull readers in and slow them down, in contrast to "macadamized" reading, which moves them along at ever-greater speeds.

How did such strategies work? Indeed, were they even possible? The discontinuous format of the periodical and newspaper had after all been shaped by the need for hasty (or lazy) reading. This variety of reading pace—rushing or ambling—was inherent in the design of the periodical, which allowed readers to select what, and how, to read. One could not compel readers to read in a particular way, but one could give advice and cultivate an experimental environment in which certain forms required slow rather than fast reading.

One important experiment in this regard related to deinstrumentalizing time, a favored Gandhian theme in which the link between speed and efficiency was questioned. In part, this unsettling of speed as an end in itself was inherent in the exchange system itself where circulation went slowly, and any news clipping was always in transit, pausing briefly between its previous origin in some other paper and its next possible destination in another. This textual environment foregrounded questions of reading and time and hence offered a propitious climate in which to cultivate certain rhythms of reading and resting, perseverance and pausing.

Extracts and essays along this theme of time and reading were hence common, with Thoreau being a favored source. "Read not the Times. Read the Eternities" ran one quotation that recognized the powerful shaping force of print media but suggested ways of resisting its vortex (June 10, 1911). The Thoreau quote cited earlier about the "macadamized" mind is worth revisiting. It begins with warnings against cluttering one's mind with trivia (or what Gandhi elsewhere called "the fetish of literacy and mundane knowledge," *CW* 10: 227).

> I believe that the mind can be profaned by the habit of attending
> to trivial things, so that all our thoughts shall be tinged with
> triviality. Our very intellect shall be macadamized, as it were,—
> its foundation broken into fragments for the wheels of travel to
> roll over; and if you shall know what makes the most durable
> pavement surpassing rolled stones, spruce blocks, and asphaltum,
> you have only to look into some of our minds which have been
> subjected to this treatment so long. (June 10, 1911)

This metaphor of "macadamized" minds presents a powerful image of what reading should *not* be. It should not be rapid, smooth, and frictionless. Instead, reading and thought should involve effort and friction, or "counter friction," as Thoreau termed it in an extract from *On the Duty of Civil Disobedience* that appeared in *Indian Opinion* on October 26, 1907. The machine of government has injustice built into it, a friction that in some cases will cause it to wear out. "But if it is of such a nature that it requires you to be an agent of that injustice to another, then, I say, break the law. Let your life be a counter friction to stop the machine" (October 26, 1907).

Against such "macadamized" newspaper surfaces, *Indian Opinion* specialized in "uneven" reading surfaces across which one could not hurtle. The article "Extracts from Balzac" strings together some two dozen bite-sized quotations. To attempt to read them continuously induces intellectual vertigo as one hurtles from one topic to another, from ideas on liberty, to comments on the nature of scandal, to observations about genius (August 12, 1911). One can only really absorb one or two quotations at a time, contemplating these and then returning later to consider another. The article defies hasty reading.

As we have seen earlier, such ethical extracts were placed between news items, a juxtaposition that asked readers to equate the two. The two forms commented on, and could be applied to, each other—the ideas of Thoreau to reports on satyagraha in Johannesburg or racial exclusion in British Columbia and Australia. A comparison between reports of such racial exclusion and Socrates's defense and death requires one to take the long view that relativized the oppressions of empire and made them seem less eternal but also brought Socrates into the present, making him a spiritual comrade and friend.

The juxtaposition of the two genres (news and ethical extract) insistently required readers to think about what constituted ethical news reporting. This was a pressing question in the context of southern Africa, where the white-run press endlessly recycled racism, hatred, stereotype, and vitriol. *Indian Opinion* routinely printed such extracts with no comment attached, a kind of textual, nonviolent, noncooperation. The July 8, 1911, edition carried an article titled "'The Sons of South Africa': A New Society," extracted from the *Rand Daily Mail*. This organization had been set up to educate South Africans about their rights and privileges in empire and "is non-racial in its policy." Yet membership was open only to those of "pure European descent." *Indian Opinion* offers no comment on the story but instead creates a border of silence around it in which the reader has to ruminate on what has been said (for further examples, see August 20, 1903; June 18, 1903; January 1, 1906; February 17, 1906).

Another strategy in cultivating a climate of considered reading was direct advice to subscribers. This forms the subject of Chapter 5, so for the moment we can note briefly that Gandhi's Gujarati columns gave frequent and detailed advice to readers on how to read in a slow and careful way. Readers were constantly encouraged to read articles repeatedly, to keep copies of the newspaper, and to make their own collection of clippings. The model of reading was one of perseverance (read the whole paper) and pausing (stop and think, come back later, reread, and reconsider).

These instructions to readers were also reproduced in the constant advice Gandhi gave to his colleagues at Phoenix, who functioned in some respects as a reading commune. In part his guidance focused on how to manage the vast amount of reading entailed in keeping on top of the exchange papers, a system of journalism dependent as much on reading as on writing. Gandhi urged the Phoenix workers to incorporate the reading of exchanges as part of a regular weekly schedule: "It will be all right if you devote Thursday to the reading of proofs and Tuesday and Wednesday exclusively to general reading and to writing Gujarati" (*CW* 5: 80). Chhaganlal must have taken on a lion's share of this work, having developed his skills reading the Gujarati exchanges in the earliest days of *Indian Opinion* on Mercury Lane. Henry Polak likewise had worked on the exchanges while employed on the *Transvaal Critic*, and much of the English gleaning and summary work must have come his way (albeit he was based in Johannesburg).

Other forms of advice pertained to the management of clippings, which formed an important strand in Phoenix life. Whether in Johannesburg, London, or Bombay, Gandhi generated swirls of letters, cuttings, and scrapbooks. Many were sent to Phoenix with instructions: some cuttings were to be translated into Gujarati and used in the newspaper (*CW* 5: 418–419); others were not destined for print but were for the edification of Phoenix residents (*CW* 6: 207). In some cases Gandhi sent entire publications that were to be read and their pictures admired, and then the periodical was to be passed on to the Indian Public Library (*CW* 6: 393, 401–402). These cuttings were accompanied by, and interspersed with, letters from Gandhi, some of which contained summaries of events that were to appear in the paper.

Other letters gave advice on what to read, and how to read it: "Please tell Maganlalbhai that I would advise him to read Emerson's essays. They can be had for nine pence in Durban.... Those essays are worth studying. He should read them, mark the important passages and then finally copy them out in a notebook" (*CW* 9: 200). A few weeks later another letter arrived: "I should like all in Phoenix to read Tolstoy's *Life and My Confessions*. Both are soul-stirring books. They can be easily read in two days. The Gujaratis

should also read Kavi's two volumes." Gandhi went on to recommend that the Gujarati speakers discuss sections of Kavi (Raychan-
dbhai Mehta, Gandhi's spiritual mentor) in the half-hour evening
services that were held daily at Phoenix (*CW* 9: 203). Just as these different forms of reading—exchanges and ethical writers—intersected
in the paper, so too did they combine in the reading life of Phoenix, making the prayer service an extension of the paper, and vice
versa.

In his interactions with Phoenix residents, Gandhi sought to
cultivate an ideal reading community in which reading would
proceed by pausing and contemplation as much as by persevering
through the heft of the weekly exchange papers.[18] At times the
practices of the Phoenix reading community manifest themselves
in the newspaper, demonstrating a model of contemplative reading, maintained in the face of deadlines and pressure to produce.
An example, again from Thoreau, illustrates the point. Under the
headline "Thoughts from Thoreau," the paper offered a selection
of quotations from *Life without Principle*.[19] The piece, a speech given
by Thoreau, initially in 1854 and then published in 1863, explores a
theme close to Gandhi's heart, namely, the state of modern civilization as one that promotes modernity without interiority. Driven
by the incessant search for profit, life becomes a series of meaningless and surface activities: work lacks inner meaning, and politics
is "superficial and inhuman." Newspapers fill people's heads with
trivia. Moving from surface to surface, we inhabit a world in which
our inner domain shrinks. Many of us "have not heard from ourselves" in a long time.

Indian Opinion ran two sets of extracts from this piece: the June
10, 1911, edition contained twelve extracts, the July 22, 1911, edition
eighteen (although nine of these are not from *Life without Principle*

and seem to have come instead from Thoreau's letters). Together the twenty-one "real" extracts, which range in length from twenty to a hundred words, make up about 20 percent of the original speech. In most cases the extracts are nonnarrative expositions, with any narrative or anecdotal "chaff" removed.

The pressure under which *Indian Opinion* was produced makes itself felt in the presentation of these citations. In the first collection of quotations, the order of extracts does not follow that of the original and, as already indicated, the second is interlarded with matter from elsewhere. Such extemporization is probably due to the pressures of printing and making text fit into the format of columns and pages. A second factor accounting for this makeshift character can be traced back to the many hands involved in the selection and production of the extracts. Gandhi probably selected most of the extracts as he had done four years earlier, in the case of *On the Duty of Civil Disobedience*. However, on July 12, 1911, Gandhi wrote to Maganlal at Phoenix and advised him to "Copy out a sentence from Thoreau occasionally and give it to Mr West for use in *Indian Opinion*" (CW 11: 465). This dating does not, however, dovetail with the actual publication of the extracts (June 10 and July 22), unless of course Maganlal was quick off the mark. Yet we do know that Gandhi urged Phoenix residents to copy out extracts from their reading, and some of these no doubt made their way into the paper.

The extracts are hence produced in a community, through a process of careful and contemplative reading, but they have to take into account deadlines and the makeshift nature of production. One has to work within the pulse of mechanical production, but one does not have to be utterly governed by it, nor take on the urgent and panicked tempo of industrial speed. We can register its mechanical rhythms differently by focusing on the pause rather than on the forward rush.

However, one may raise questions around the ethics of reading only 20 percent of the original. Is this not playing fast and loose with the integrity of texts? Is this fraudulent and superficial reading? Not if we consider this text-making in the context of *Indian Opinion* and its methods of writing-as-editing in which we expect texts to come to us already mediated. "Thoughts from Thoreau" announces itself as a "made-up" text, selected from elsewhere. As we make our own choice from the selection, we reenact the process of textual production in slow motion. It is as if we are constructing our own miniature handmade newspaper, reprising an ideal version of the Gandhian newspaper—hand-selected, read with care, and educating a reader to become an independent interpreter and selector.

An anecdote from Tolstoy Farm, which was established in 1910 outside Johannesburg as a headquarters and ashram for satyagrahis and their families, illustrates these themes well. From time to time some Tolstoyans, as they called themselves, would elect to walk the twenty-odd miles from the farm to Johannesburg. The fastest was Hermann Kallenbach, a Gandhian supporter who had purchased the property: leaving at 3:00 A.M., he could cover the distance in just under five hours. On one occasion, when Johannesburg residents saw Kallenbach striding with such speed and purpose, they assumed he was taking part in a racing event. Yet the walking was in fact almost the opposite and constituted but one measure of disciplined living practiced by Tolstoy residents (May 11, 1912).

This story of speed-that-is-not-speed, of racing that becomes its opposite, provides a useful analogy to *Indian Opinion,* a newspaper-that-is-not-a-newspaper. To the casual observer, *Indian Opinion* appears to be doing the work of any ordinary periodical, namely, promoting hasty and discontinuous reading. Yet *Indian Opinion* strove to turn such a format against itself by organizing reading around pausing and perseverance. Instead of seeing the gaps between stories as something to be hastened over, these become moments for reflection, and for thinking about how to link the bits one has read.

4

Binding Pamphlets,
Summarizing India

Between 1903 and 1914 the International Printing Press (IPP) published some thirty pamphlets, mostly taken from popular or significant articles in *Indian Opinion*. Writing in 1913, Gandhi described these as the most important part of the paper: "Our purpose is to publish, from time to time, articles of permanent value so that readers who like to preserve copies can later have them bound into a volume" (*CW* 12: 360).

So central were these publications that at times the paper was envisaged as serving the pamphlets rather than vice versa. As part of the same downsize, the paper shifted from three to two columns. This change aimed to improve the appearance of the newspaper and make matters more convenient "if the articles had to be published in book form" (*CW* 12: 360).

The technicalities inherent in this statement remain unclear: it implies either that type could be spared, and the forms left intact until the pamphlet could be issued, or that stereotypes of them would be made. The chronic shortage of type, especially in the case of Gujarati, rules out the first option, while the cost of stereotyping, plus the fact that it is never mentioned in the extensive correspondence between Gandhi and the printing personnel at Phoenix, makes the second unlikely. Furthermore, the physical size and appearance of the pamphlets differ from their original form in the newspaper, indicating that they were all composed anew.

Yet even if not in the form set out earlier, the pamphlets did shape the newspaper in other ways. In 1912 Gandhi gave up virtually all advertising for the paper, arguing that the time spent in soliciting advertisements could be better devoted to the production of booklets.[1] It is as if the pamphlets replace the advertisements, the hasty tempos of the market giving way to the slow and more durable reading of the Gandhian "book."

The pamphlets have become one of the paper's most enduring legacies. Titles like *Hind Swaraj* (1909) rose up from the pages of an obscure newspaper to become world famous. Others, like *Sarvodaya* (1908), Gandhi's adaptation of Ruskin's *Unto This Last*, and his abridged account of Socrates's defense, *The Story of a Soldier of Truth* (1908), feature centrally in discussions of Gandhian ideas.

These texts are the choice kernels winnowed from the paper, leaving a chaff of lesser pamphlets that have blown away. Apart from advertisements in the paper or the booklets being banned by the Government of India (the fate of four Gujarati pamphlets in 1910), there is little evidence of these tracts ever existing. Such a fate for a pamphlet is not unusual. As Peter Stallybrass points out, while books are designed for survival, pamphlets perish along with the ephemera that make up the bulk of a printer's business.[2] One way in which pamphlets might have survived is via copyright mechanisms designed to direct publications to deposit libraries. Yet in at least two cases, *Hind Swaraj* and Tolstoy's *Letter to a Hindoo*, the pamphlets were explicitly not copyrighted.

Consequently, it is difficult to know the total number of pamphlets. Working from the English sections of the paper and translations of the Gujarati book columns, as well as mentions of the pamphlets in other contexts (like banning orders or from Indian periodicals that received copies), it would seem that there were six pamphlets in English and some two dozen in Gujarati (see the Appendix).

Many hands produced the English pamphlets: notable authors were Gandhi, Polak, and Tolstoy (in translation). The Gujarati tracts emanate mainly from Gandhi and include translations of laws and ordinances, translations of ethical writings, and biographies. There is also a collection of poems sent in by readers. Apart from *Hind Swaraj*, the leaflets do not constitute original "writing" but, rather, consist of abridgments, retellings, reviews, and translations.

Binding Together

This chapter takes seriously Gandhi's injunction to bind the pamphlets—or at least some of them—together (as the book pages of *Indian Opinion* did on a weekly basis, advertising a "bookshelf" of its publications). In giving this advice, it is as if he wants the best of both worlds—the provisionality of the pamphlet and the permanence of the bound book. The material form of such a volume would keep this duality alive: each pamphlet produced by *Indian Opinion* was a different page size, and even if bound together the distinctiveness of each item would be manifest.

This duality of book and pamphlet furnishes the method for this chapter. Considering the first part of the equation, we ask whether one can create a "book" of these pamphlets. Could we treat them as fascicles, parts of a book published separately in installments? A fascicle is also a close cluster, whether of flowers and leaves or nerve and muscle fibers. These multiple meanings of the word serve our purposes well. While never intended to be one monumental book, the pamphlets do constitute a bundle, a thickening of texts whose juxtapositions are as important as their component parts. In keeping with the spirit of exchange journalism, the omnibus configures an unusual cast of characters and intellectual traditions: Socrates; the Egyptian nationalist Mustafa Kamal Pasha; Ruskin; Annie Besant; the jurist and politician Syed Ameer Ali; and J. L. P. Erasmus, a Boer commandant (who provides an English retelling of the Hindu epics). What kinds of intellectual histories emerge if these are bundled together? Likewise, what will emerge from the constituent parts themselves, mostly compacted miniatures: retellings, reviews, extracts, abridgments, abridged translations?

Summary India

This theme of the miniature and the summary speaks as well to Gandhi's larger project in South Africa, namely, constructing a "summary India." With groups from across India thrown together in South Africa, it became possible to grasp India in a way that was not possible on the vast subcontinent itself. The newspaper itself sought to fashion a community of British Indians (and imperial citizens) rather than readers who saw themselves as "Tamils or Calcutta men, Mohamedans or Hindus, Brahmans or Banyas" (*Indian Opinion,* August 20, 1903). This definition of India necessarily depended as well on the broader contours of southern African societies: white settler nationalists, African chiefdoms, Boer republics, and migrant workers, both African and European, all of whom provided different kinds of boundaries for the idea of India. In this multiply diasporic context, Gandhi sought to forge a durable definition of who could be Indian, both outside and inside India.

Combining different communities into a nation located outside itself was necessarily a demanding utopian task. On the one hand, potential readers had to be addressed in linguistic compartments if one wished to draw them together in unity. Yet, on the other hand, the nature of that unity was uncertain, since the idea of India on the subcontinent itself had still to be converted into a nationalist idiom. There was hence no ready-made original to which one could make recourse. Through its initial four-language policy, *Indian Opinion* attempted, visually and linguistically, to "collate" different communities into one space. When, against Gandhi's wishes, the newspaper discontinued Tamil and Hindi early in 1906, he tried to keep some vestige of this version of the paper as India alive, suggesting that the idle Hindi type be used to produce a pamphlet version of the first canto of the "Hindu Bible," namely, Tulsidas's Ramayana (*CW* 7: 60).

As extensions of the paper, the pamphlets were hence tasked with taking forward experiments in audience-making, which were inseparable from the process of collating different configurations of how to be Indian. Each pamphlet acted as a hypothesis, presupposing various configurations of Indian-ness. Nation, empire, religion, region, civilization, race, language, caste, body, soul—which of these could one dispense with and which not? The pamphlets juggled these categories, adding and subtracting them in different combinations and building up a portfolio of possible ways of being Indian. The results of these experimental conversations culminated in *Hind Swaraj*, which offers an elegantly minimal definition of Indian-ness, as the next chapter argues.

In pursuing this task, some pamphlets addressed (or attempted to address) specific subaudiences, while others spoke to some broader ideal of British India. The abridged form of the pamphlet lent itself well to this latter utopian task, being able to imply a distant original yet located far enough from it to be free to experiment. As a compendium of miniature epics, edited extracts, and summarized translations, the pamphlets usefully appear to be about scale, and hence to bear some relationship to an original model. But even where there is nominally an original (as in the pamphlet versions of the Hindu epics, or in cases of summarized translations), it is so remote, or is treated with such dispatch, that it exists only in the faintest outline, opening up an experimental environment that these pamphlets embody. They are acts of radical collating and editing, not simply in the sense of paring down an original but in the older sense of putting forth or publishing a text.

Interestingly, questions of caste are never addressed directly in these pamphlets or, indeed, in *Hind Swaraj*. One reason for this silence, as Joseph Lelyveld suggests, is that caste was seen as an internal issue to be discussed among Indians and was not to be aired too frequently in public lest colonial society seized on it as a further stick with which to beat and belittle the Indian community.[3]

Miniature Epics

In its early years *Indian Opinion* produced five versions of the Hindu epics: two Gitas, two Ramayanas, and one Mahabharata. All were in English, except for one that consisted of the first canto of the Tulsidas Ramayana in Hindi. The epics were distributed in and outside the paper. One was printed first as a series in the paper and then as a booklet; two were published only as pamphlets. The remaining two appeared only in the paper, although it probably had been intended to give them a subsequent pamphlet existence.

The movement of these versions in and out of the paper captures in miniature the larger flow of these narrative traditions in and around the old and new South Asian diasporas. Paula Richman has examined aspects of this movement, looking at Ramlilas in the post–1960s South Asian diaspora to Britain. Discussing the old indentured diaspora, John Kelly has demonstrated how laborers in Fiji used aspects of the Ramayana tradition to give their own suffering epic proportions.[4] In Natal similar patterns emerged: indentured plantation workers placed themselves as characters in the story of Ram, often as Hanuman clearing the way for civilization, a lesser known variant in the vast sea of what Philip Lutgendorf calls Hanumayana.[5]

The *Indian Opinion* epics departed from these patterns: they were not aimed at indentured workers and spoke to rather different kinds of diasporic audiences. We provide a brief account of each pamphlet before turning to questions of audience and ideas of Indian-ness.

Besant's Gita

"Special South African Edition: Mrs Besant's Translation of the 'Bhagavad Gita' with Selections in Sanskrit and Sanskrit Alphabet with English Pronunciation for Students. Full-page Portrait of Mrs Besant. One Shilling"—thus read an advertisement in the December 10, 1904, edition of *Indian Opinion*.

No versions of this pamphlet survive, but the circumstances of its production emerge from a mollifying letter Gandhi wrote to an indignant Annie Besant who objected to her translation being reproduced without permission and her portrait being included in a sacred work.

> A gentleman offered to have a translation of the Bhagawad Gita printed for distribution among Hindu boys and other[s] if the Managers printed it at cost price. He was also in a hurry. Reprint of your translation was suggested. The matter was referred to me and, after much careful thought, as there was no time left for reference to you, I advised that your translation might be printed for circulation in South Africa. I felt that the motive of the management was pure, and that when the circumstances, under which the edition was published, were brought to your notice, you would overlook any apparent impropriety. (CW 4: 271–272)

The letter provided figures on print run and sales: 1,000 copies had been printed, and 200 remained, being sold or distributed at the rate of five per month, and "then only among real inquirers" (ibid., 272). While we do not know who "the Hindu gentleman" was, he certainly formed part of a growing body concerned about religious and linguistic attrition in the diaspora.

Beyond Boer Orientalism

A page before the advertisement for Besant's Gita, one encounters the headline: "The Psychology of the Bhagavad Gita: A Lecture Delivered before the Transvaal Philosophical Society: By Mr JLP Erasmus, Ex-Commandant of the Boers" (December 10, 1904).

Who was Erasmus, and how did he come to be writing for *Indian Opinion*? Born in the Cape Colony in 1858, Erasmus had moved north and ended up as an attorney in the Transvaal republic. At the outbreak of the Anglo-Boer War in 1899, he headed up a Boer

commando until captured by the British in March 1901. Expecting
to win the war in three months, the British had made little provi-
sion for holding prisoners of war (POWs). As the war dragged on
and POW numbers mounted, the British turned to a well-tried solution: the empire as gulag. Boer POWs were initially shipped to
Ceylon, St. Helena, and Bermuda. When these island prisons filled
up, the British leaned on a reluctant Government of India to take
Boer POWs. Erasmus was one of nine thousand POWs interned
across British India between 1901 and 1903, when the last POWs
were finally shipped home.[6]

After being held in Cape Town, Erasmus was transported to
Bombay in October 1901 and then on to Shahjehanpur before being
interned in Fort Govindgarh, just outside Amritsar. In an article in
Indian Opinion, Erasmus presents his time in India as an education
in understanding hierarchy (November 5, 1904). He leaves South
Africa with crude orientalist ideas and itches to see fakirs perform-
ing mind-boggling miracles. When Erasmus arrives in the intern-
ment camp, he mistakes Sudras for sannyasis but is put right by
the divisional sergeant who starts to educate him about caste. At
the end of the war he is released, but while waiting to return home
he is granted daily parole and befriends a lawyer, Saran Dass Cho-
pra, who introduces him to his circle of friends who usher him up
the caste hierarchy "until at last I had the honour of meeting
gentlemen of the Brahman caste." Under their influence, he can
distance himself from his earlier crass orientalism, having replaced
it with a form of mutual recognition between colonized elites who
educate each other about their respective forms of social distinction. These Erasmus reproduces: an article on Indian history
drawn from R. C. Dutt's *Epochs of Indian History* produces a Hindu-

centric, anti-Muslim overview of the subcontinent in which caste represents a crowning achievement.

During his time in Amritsar, Erasmus acquired a small library of books on India provided by his Brahmin friends. On his return to Johannesburg, he put these to good use and delivered a series of lectures to both the Transvaal Philosophical Society and the Transvaal Theosophical Society on the Gita, the Ramayana, and the Mahabharata, drawn from English translations: Sastri for the Gita and Romesh Chunder Dutt's versified abridgment of the Ramayana and Mahabharata, as well as Manmatha Dutt's translation for the latter.

Gandhi and Polak both had links to these societies, and one or both of them may have attended the events and approached Erasmus to place his lectures in *Indian Opinion*. Across a period of eighteen months, from November 1904 to May 1906, the paper featured twenty articles by the Boer commandant.

On January 28, 1905, *Indian Opinion* profiled Erasmus as the ideal contributor, holding him up as an example to other European readers. Despite being a Boer (a group possessed by "more than usual colour-madness"), his lectures show how "a prominent member of that nation can become sympathetic by the simple process of learning the truth." Erasmus's "excellent resumé" showed that "India had a high civilization, when Europe was sunk in barbarism."

With regard to the structure of the booklets, Erasmus's pamphlets necessarily offer a summary of a summary of a translation of a translation. His foreword to the Ramayana draws substantially on Dutt's epilogue, reproducing nuggets with minimal paraphrase. With regard to the narrative structure itself, Erasmus positions himself as a textual choreographer, now narrating segments of story, now ushering in Dutt's voice, and now marshaling lines of direct quotation. The narrative voice appears as one that stands in among a cloud of quotation.

With regard to audience, Erasmus's talks address a group that is implicitly Christian, European, and conversant with Western philosophy, interlarding his accounts of the Gita with comparisons to Trinitarian theology and invoking Kant and Plato on the soul.

Tulsidas's Ramayana

In October 1908 Gandhi wrote an endorsement for the upcoming publication of a portion of the Tulsidas Ramayana in Hindi. He opened with the theme of religious attrition in the diaspora:

> These days India's sons go abroad in large numbers. In a foreign country, not everyone can be conscious always of his particular religion. This is especially so in the case of the Hindus. The present writer is of the view that it is the duty not merely of the Hindus alone but also of all Indians to acquaint themselves with the essentials of Hinduism in its common form. (CW 9: 202)

Gandhi endorses Tulsidas's work as reflecting "most vividly . . . the general spirit of Hinduism." It is almost as good as Valmiki's Sanskrit original and appears not to be a translation at all: "His devotion to God was so profound that instead of translating, he poured forth his own heart" (ibid.).

Those settled in "foreign lands" often do not have time to read the whole work. An abridgment may hence be beneficial not to replace the original but to entice readers to consult the whole work. "The abridgement does not leave out any portions of the main narrative. But interpolations, long descriptions and some portion from the subsidiary parts have been omitted" (CW 9: 203).

> We wish that every Indian goes devoutly through the summary which we are placing before the public, reflect[s] over it, and assimilate[s] the ethical principles so vividly set out in it. We shall consider our effort to have been duly rewarded if this abridged Ramayana is read in every Indian home in the evenings and during the periods of leisure at other times. (ibid.)

The remaining cantos were to follow (something that appears not to have eventuated), and readers were urged to collect them and bind them into one volume. The price was kept low to place the book "within the reach of every Indian" (ibid.).

Getting Tulsidas's Ramayana into print was a project that mattered to Gandhi, as evidenced by his pressure on Chhaganlal to complete the assignment (CW 7: 60). As we have seen, the idea took shape in 1906 when the paper dropped its Hindi and Tamil section. Having lost one way of embodying "India" through a newspaper in four languages, Gandhi turned the Hindi type toward another, namely, to produce what many regarded as "the Hindu Bible." What did this say about the configurations of audiences for these epics and their alignment to "India"?

Epics and Audiences

The intended audiences for these epics were threefold: Europeans and elite Indians for Erasmus's versions; "colonial-born" Indians for the Besant's Gita; and Hindi speakers for Tulsidas. While we do not have much evidence, the two latter pamphlets seem to have found their intended audiences. Besant's Gita certainly proved popular and was still being sold in the 1920s, presumably largely to "colonial-born" Indians to whom English was an ever-more-central language (*Indian Opinion,* December 29, 1922). The Tulsidas Ramayana likewise proved to be a perennial and was still on sale in 1913, acquired by Hindi speakers, most of whom spoke Bhojpuri-Hindi (*Indian Opinion,* April 12, 1913).[7] Most buyers were probably literate, although there could have been those of uncertain literacy who obtained it as a talismanic object that symbolized Hindu India.

Erasmus's Ramayana, in contrast, reached an audience different from what had initially been intended. Placed in the English columns, its imagined readers were those of the English section as a whole: European sympathizers in South Africa and further afield, supporters in India, and local members of the Indian community who read English.[8] As an aside, the English section would also have been followed by two further groups: journalists on other papers and colonial officials keeping their collective finger on the pulse of political developments—the colonial state was a regular subscriber to, and careful reader of, the journal.[9]

Yet this Ramayana found relatively few readers among these groups. Some European sympathizers might have read the epics to educate themselves, but the other constituencies could hardly have paid much attention. Where readers did, they were largely irritated at Erasmus's ham-handed efforts. One subscriber using the pen name Kamadhuk (cow of plenty) took issue with Erasmus's Christian interpolations and his lack of Sanskrit (February 4, 1905). Another, J. E. Done, did not specifically engage with Erasmus but wrote a piece on Indian history in part to show up his more in-depth knowledge in comparison with Erasmus (February 10, 1906).

Yet once released from the paper and put into pamphlet form, Erasmus's Ramayana found a new and enthusiastic audience among "colonial-born" readers.[10] In fact, the tract proved to be the newspaper's most constant seller and was still being advertised in the 1930s (September 9, 1932). Its appeal lay in the fact that it was short and easy to read: about six thousand words spread across some forty generously spaced pages. A less tangible aspect of its attraction may be the fact that Erasmus, like many "colonial-born" Indians, spoke no Indian languages. The text comprised bits and pieces of translated "tradition" and hence mirrored the relationship of "colonial-born" Indians to most matters Indian.

The miniature epics played a role in clarifying different audiences around the idea of India, especially since the epics, or at least their distant originals, could stand in for the idea of India itself. Yet freed from that original by their hyperexcised forms, they enabled more experimental notions of Indian-ness in which one could be Indian without speaking an Indian language. There was of course an implicit proviso that one be of Indian descent, but the Erasmus case opens up a radical possibility of being Indian solely through apprenticing oneself to its civilizational traditions. This interpretation may sound somewhat far-fetched, but as will be shown in Chapter 5 it was a possibility that made its way into *Hind Swaraj*. In the Erasmus hypothesis, whiteness, Christianity, and lack of an Indian language would not be absolute barriers to entry. One's claims of belonging might be weaker than some, but one could in theory find a place, provided the definition of "India" was capacious enough. However, there was one nonnegotiable condition of entry: "civilization." Those ostensibly without it—like Africans—could not gain admission.

The epic pamphlets functioned rather like open invitations that produced surprising results: some expected groups (like the Hindi speakers) showed up, others (like the European readers) did not, while unexpected groups (like the "colonial-born" Indians) arrived in numbers. Still others, like the ideal British Indian imperial citizens, could not take up the invitation, as they hardly existed.

These different audiences also chafed against each other, representing larger fault lines in the South African Indian community itself, in this case between "colonial-born" Indians and those from the mainland. While the "colonial born" were certainly addressed in the paper itself and were imagined as one of its outlier constituencies, they were never enthusiastic readers of a paper that patronized and upbraided them, and that offered little in the way of popular culture and sport.[11] Gandhi himself regarded "colonial-born" Indians ambivalently. He worked closely with their leaders in the satyagraha campaigns but expressed skepticism about their rank and file. In the midst of a repatriation crisis in the mid-1920s in which "colonial-born" Indians were offered cash inducements by the South African state to return to the subcontinent where they struggled to fit back into a caste-dominated society, Gandhi noted: "The men are neither Indian nor Colonial. They have no Indian culture in the foreign lands they go to, save what they pick up from their uncultured half-dis-Indianized parents" (CW 36: 285–286).[12] Polak shared similar sentiments: "There they were [in South Africa], helpless in the midst of an alien population, whose civilization was incomprehensible to their generally limited intellects, and whose mental attitude was coloured by the long contact with a savage race of aboriginals."[13] "Colonial-born" Indians constituted suspect subjects of the virtual nation who needed to be tutored and guided to full citizenship.

Unsurprisingly, in 1908 a newspaper tailored more to the taste of a "colonial-born" constituency, *African Chronicle*, was started by P. S. Aiyar, and it probably drew away any lingering "colonial-born" readers of *Indian Opinion*. Yet as we have seen, the "colonial-born" constituency continued to buy the Erasmus Ramayana and Besant Gita. The pamphlets managed to retain these readers as a peripheral audience, and hence they kept alive some vestige of the ideal of a pan-Indian audience.

However, this pan-Indian ideal of the paper continued to fragment, as other groups also broke away to form their own newspapers. Two papers aimed at Muslim readers—*Al-Islam* started in 1907 and *Indian Views* in 1914—further splintered the readership although certainly did not drive away all Muslim subscribers. Those with the means probably bought all the papers they could (*CW* 12: 367), and there was something of a friendly rivalry between the papers, which jousted with each other in their columns (*CW* 7: 284) but also assisted each other: *African Chronicle* supported the satyagraha campaign and translated one of Gandhi's speeches into Tamil in 1909.[14]

In a community as diverse as Indians in South Africa, such fragmentation on linguistic, caste, and religious grounds was hardly surprising. One response from Gandhi was to persevere with a utopian ideal of creating a miniature and united India. The columns of *Indian Opinion* continued to be written as though all South African Indians could be addressed via Gandhi's column, which routinely hailed all Indians as though they understood Gujarati (a topic I touch upon in the next chapter). Indeed, one strand in Gandhi's thinking about translation, namely, that emotion could trump language, illuminates further this will to communicate despite the barriers of multilingual life.[15] Yet the difficulties of trying to forge a polyglot, polyreligious, and polycaste nation out of a set of determinedly sectional units were overwhelming. One response on Gandhi's part was to radically recast the idea of the nation and sovereignty itself. What if a nation was less an abstract cloak thrown over a congeries of people and was more an ideal that had to be fostered from the ground up, one person at a time?

An analysis of a further tranche of pamphlets illustrates this trajectory in Gandhi's thinking. We begin with some pamphlets aimed at Muslim readers and then turn to two exhortative pamphlets, each stepping-stones on the road to defining the satyagrahi as the locus of the nation.

Muslim Pamphlets

The English translations of the Hindu epics all appeared between late 1904 and 1906. In a situation where initially the majority of Gandhi's support both for satyagraha and the newspaper came from Muslim merchants, this absence of any pamphlet aimed at Muslims must have been apparent. Writing in the Gujarati columns on January 5, 1907, Gandhi noted his intention to put matters right:

> Considering that a large number of our readers are Muslim, we are thinking of publishing a translation of the celebrated Mr. Justice Ameer Ali's book on Islam [*The Spirit of Islam*] . . . Justice Ameer Ali has given us the permission to translate it. The consent of the publishers is yet to be had. If that is also received and if the idea is favoured by our readers and they are prepared to encourage us, we intend to publish a translation of The Spirit of Islam in book form. [It] has won fame throughout the world and deserves to be read by every Muslim, indeed, by every Indian. (CW 6: 209)

Two Gujarati pamphlets containing speeches and articles by Ameer Ali did subsequently appear, although the sequence is not entirely clear: "There are difficulties in the way," Gandhi noted (CW 7: 16). Possibly during this period of delay, Gandhi turned his attention elsewhere, and on June 22, 1907, announced in the Gujarati columns that *Indian Opinion* would carry an abbreviated translation of Washington Irving's *Life of the Prophet*.

> It will always be our aim to bring about and preserve unity between Hindus and Muslims. One of the ways of achieving this is to acquaint each with whatever is good in the other. Moreover, when occasion requires, Hindus and Muslims should serve each other without any reserve. The series that we are commencing is intended to serve both these aims.... Most Hindus are ignorant of the career of the Prophet. Most Muslims are ignorant of the researches made by Englishmen and of what they write about the Prophet. The history by Washington is likely to be of benefit to both of these classes [of reader]. (*CW* 7: 16)

The first episode appeared on June 22, 1907, but soon ran into trouble. Some readers complained about the way in which the Prophet's marriage was portrayed. There were also intimations that readers were irritated at having a non-Muslim and a white discoursing about Islam. Gandhi, in contrast, insisted it was important to know what whites wrote about Islam, a move that implicitly equated Irving on Islam with Erasmus on Hinduism (*CW* 7: 172–173).

While some readers wanted the series to carry on, it was nonetheless discontinued. Gandhi urged those in favor of the series to write to the paper: "If many readers express the desire, we shall try to meet the wishes of such devout men by bringing it out separately in book-form when convenient to the Press" (*CW* 7: 173). Whether only a few readers responded or because of the history of the controversy, this pamphlet never saw the light of day, although *Indian Opinion* did advertise the English version of the text in its book columns (December 16, 1914).

In a context where Gandhi sought to invent India in miniature, he strove to represent all categories in the paper, with the phrase "Hindu, Muslim, Parsee, Christian" a constant refrain in its columns. In 1905 he produced a series of biographies comprising about a dozen stories of heroes, whose example readers were intended to emulate, and which he urged readers to gather together in their own homemade pamphlets. Most figures in the series were Europeans, but three were Indian and formed a semirepresentative trio: an Urdu Muslim poet (Hali, whose work deals with "such useful themes as the duty of Muslims in the present age," CW 4: 446); a South Indian administrative reformer (Madhav Rao, the "Pericles" of Madras, CW 4: 461); and a leading figure of the Bengal Renaissance and philanthropic reformer (Ishwarchandra Vidyasagar, CW 4: 411–414).

Elsewhere Gandhi sought to profile prominent Muslim figures. Two pamphlets—one a biography, the other a speech—dealt with the Egyptian nationalist Mustafa Kamal Pasha, although he is only incidentally Muslim, his profile as an Egyptian nationalist and editor being more prominent.

Taken together, these pamphlets aimed at Muslim readers met with mixed success. The Syed Ameer Ali pamphlets proved to be popular and consistent sellers (Indian Opinion, December 28, 1912), while the projected Irving pamphlet proved to be a nonstarter. The tensions inherent in this situation point to the difficulties of establishing a "portfolio" nation. In a situation where Muslims constituted the majority of the Indian elite in southern Africa, this minoritization must have alienated some readers, who, in later years, could transfer their allegiances to Al-Islam and Indian Views. A portfolio of equal miniatures was not equally attractive to all.

Polak's Pamphlet: *A Book and Its Misnomer*

The ideal of empire held that there would be rights for all civilized men. In this equation Europeans and Indians (but not Africans) could and should be equal imperial citizens. Erasmus and the Europeans who supported Gandhi demonstrated such principles. But what about the whites who behaved in an uncivilized way like the lynch mob on the foreshore? Polak addressed this issue in a 1907 pamphlet, *A Book and Its Misnomer*, which advanced the argument that white racism had to be understood as a symptom of Western industrial capitalism.[16] These ideas form part of an ongoing conversation between Gandhi and Polak and issue ultimately in *Hind Swaraj*.[17]

The booklet started life as a review of L[awrence] E[lwin] Neame, *The Asiatic Danger in the Colonies*, which ran across four issues in May and June 1907 (May 11, 18, and 25, and June 1, 1907).[18] Neame was a prominent Johannesburg journalist, and subsequently editor, then working on the *Rand Daily Mail*, having come from Bristol via India where he spent three years at the *Times of India*. In the small intellectual circles of Johannesburg, he was well known to both Gandhi and Polak.[19]

Neame's book had first appeared as a series of separate articles in a number of periodicals, namely, *Empire Review, Daily Mail, Pall Mall Gazette, Pioneer* (Allahabad), and the *Rand Daily Mail*.[20] Neame was a keen reader of *Indian Opinion* and drew upon it for data for his articles. He also makes references to Polak's own writings in journals like the *Empire Review*. In the enmeshments of exchange journalism, the paper reprinted some of Neame's pieces, which contain references to *Indian Opinion*, duplicating in its own columns its reproduction elsewhere (April 21, 1906; May 12 and 19, 1906).

Neame's book addressed itself to "a World problem": the Asiatic "invasion" of the world in which a "diminishing white population is condemned to a hopeless struggle for bare existence against an ever growing mass of Asiatics."[21] Polak's review focuses on the "Caucasian bias," which in Neame's argument is based on a belief that there is "an unutterable, indescribable, impalpable something that permanently distinguishes him unfavourably from his non-Asiatic brother." The review unmasks the fear that underwrites this position, "lest the phantom of an alleged superiority should be discovered and exposed to public derision—a terror lest the windy dummy of inflated self-importance be pricked."[22]

In an important move, he links this fear to how Neame defines life itself, namely, as the desire for bodily comfort and endless accumulation. Polak links this view to the broader pathologies of the West: materialistic, militarized, blinded by science, and driven by lust of domination. What, then, should life be?—"a wise contentment or a fevered avidity? A generous use or a foolish accumulation? The control of desire or its augmentation?"[23] The pamphlet expresses the hope that the demons of the West will be humanized and spiritualized by the East. "To-day, the West craves for mechanical and material perfection. Our prayer is that the East may be spared a like disaster."[24]

In a utopian crescendo, the pamphlet imagines a world humanized by the values of the East: "In that day, men will know no differences of race and colour. All will be giants, strong to defend and protect the right; when self-help will wait upon self-sacrifice, and self-devotion will be the handmaiden of self-conquest." In this utopian order even Neame, his awful book forgiven, can find a place: "And who knows but that, in that day, a regenerate and rightly-inspired Neame will have realized the beneficence of the great Power that has mercifully consigned to a well-merited oblivion his fatal fruit, turned ashes, of his ancient rashness and immaturity!"[25]

Ruskinian Body, Gandhian Soul

Polak's pamphlet bears a strong Gandhian imprimatur, giving expression to Gandhi's core ideas that autonomy resides in rule of the self. Gandhi had long been experimenting with these ideas, and in a burst of pamphlet writing between 1907 and 1909, he refined these notions, giving them the suppleness that would result in *Hind Swaraj*.

His first pamphlet during this period was *For Passive Resisters*, which reproduced the selections from Thoreau's *On the Duty of Civil Disobedience* that had appeared earlier in the English columns of October 26, 1907. Then followed four pamphlets initially published in the Gujarati columns: *The Story of a Soldier of Truth*, an abridged translation of Plato's *Apology*; *Sarvodaya (Unto This Last)*; and the two pamphlets on Mustafa Kamal Pasha mentioned earlier.[26] A fifth pamphlet took the form of a Gujarati translation of M. S. Maurice's *Ethics of Passive Resistance*, the winning entry for an essay competition. Joseph Doke, who was then acting editor, thought the essay of not much value, but Gandhi presumably thought otherwise, taking the time to translate it (*CW* 8: 270–271).

In their newspaper life, these texts were intertwined: the Socrates series ran back-to-back with *Sarvodaya*, together covering three months' worth of weekly translations (an extraordinary achievement given the other demands that Gandhi faced). Interspersed with Socrates are the four installments of the exemplary life of Mustafa Kamal Pasha. These physical interweavings become intellectual ones, with Ruskin the English antimodernist becoming an intellectual descendant of the ancient philosopher: "It can be argued that Ruskin's ideas are an elaboration of Socrates's. Ruskin has described vividly how one who wants to live by Socrates's ideas should acquit himself in the different vocations" (*CW* 8: 317). Plaited together, these pieces distribute satyagraha as a practice across time and space—ancient Athens, contemporary Egypt, industrial England—and form part of Gandhi's proclivity to universalize satyagraha, variously claiming suffragettes and African women antipass campaigners as passive resisters (July 22, 1911; April 1, 1914).

In refining his ideas on satyagraha in these pamphlets, Gandhi continued to work toward a portable definition of Indian-ness, which ultimately he found to reside in the idea of the soul, in Uday Mehta's words "a state of inwardness as the ground for moral and political action."[27] A consideration of *Sarvodaya,* Ruskin's critique of political economy provides one route into considering the unfolding of this idea. Toward this end we compare Ruskin's original with a back translation in the *Collected Works.*

The preface to the nine-part series stresses that this is less a translation than a summary that seeks to give the substance of the book for "Indians who do not know English" (*CW* 8: 317). The summary is short and roughly halves the original length of some twenty thousand words. The pace of translation also varies: the first half of the original occupies 90 percent of Gandhi's adaptation, while the last 10 percent gallops through the second half.

Some of the cuts that Gandhi makes are predictable. As he himself explains, he removed "Biblical allusions etc.," and he changed the title, which is drawn from the parable of the laborers in the vineyard in Matthew 20:14, "I will give unto this last, even as unto thee" (*CW* 8: 318). Other cuts include classical references, imperially minded passages on Indian barbarism, and abstruse discussions of theorists like Mill and Ricardo that fall in the second half of the original.[28] The translation has also been lightly "indigenized"— shillings become rupees, examples of commercial perfidy are drawn from the local merchant world, and some depictions of labor conditions appear closer to those of the diasporic worker in a merchant family than the feudal and industrial conditions that Ruskin compares (*CW* 8: 368, 419, 361).

Yet over and above these predictable ellipses and changes there is another order of excision worth noting. Ruskin's text generally gives extended passages of action to illustrate his more abstract claims. So, for example, in arguing for a fixed wage and guaranteed employment, the text provides examples of the deleteriousness of the current system, which throws "both wages and trade into the form of a lottery."[29] He follows this up with exemplary passages like the following:

> The masters cannot bear to let any opportunity of gain escape them, and frantically rush at every gap and breach in the walls of Fortune, raging to be rich, and affronting, with impatient covetousness, every risk of ruin, while the men prefer three days of violent labour, and three days of drunkenness, to six days of moderate work.[30]

Such passages generally disappear from Gandhi's adaptation, a peculiar omission on the face of it since he was ever a lover of the concrete above the abstract. In part, passages like these fell victim to the sheer compression involved and in the haste with which the translations had to be produced.

However, Catherine Gallagher's ingenious reading of *Unto This Last* provides us with another way of thinking about these omissions. As she demonstrates, even though a fierce critic of orthodox political economists, Ruskin shared with them the idea of wealth being rooted in, and measured by, the state of the body. His disagreement with political economy lay in the fact that it had "abstracted value, severed it from flesh and blood."[31] While Ruskin casts his ultimate objective as reinstating the soul into a field that has defined the body as a machine, he nonetheless, as Gallagher argues, is forced onto a terrain where a labor theory of value reigns supreme, and in which the "laboring body is the source of all value" with economic exchange proceeding "from flesh back to flesh."[32]

Value resided in bodily well-being or, as a famous passage from *Unto This Last* argues,

> In fact, it may be discovered that the true veins of wealth are purple—and not in Rock but in Flesh—perhaps even that the final outcome and consummation of all wealth is in the producing [of] as many as possible full-breathed, bright-eyed, and happy-hearted human creatures. Our modern wealth, I think, has rather a tendency the other way; most political economists appearing to consider multitudes of human creatures not conducive to wealth, or at best conducive to it only by remaining in a dim-eyed and narrow-chested state of being.[33]

Gandhi's rendition of this paragraph is as follows:

> It may even appear on a fuller consideration that the persons themselves constitute the wealth, not gold and silver. We must search for wealth not in the bowels of the earth, but in the hearts of men. If this is correct, the true law of economics is that men must be maintained in the best possible health, both of body and mind, and in the highest state of honour. (*CW* 8: 406)

Gandhi's insertions are clear: wealth resides not only in the body but "in the hearts of men," and the true law of value lies in "the highest state of honour."

While Gandhi and Ruskin share many ideas in common, this translation points to a subtle parting of the ways, one pulled despite himself into a field of discourse where bodily well-being forms the basis of value, the other pulled toward an order where the soul underwrites ideas of value, autonomy, and sovereignty. The translation itself mirrors this separation of ways. The conclusion abandons Ruskin entirely and turns to consider questions of what constitutes true self-rule, providing us with a miniature rehearsal of *Hind Swaraj*. Gandhi first rejects ideas of self-rule as simply a political arrangement. Natal and Transvaal, as colonies with responsible government, possess self-rule but have created abominable societies based on the maintenance of bodily welfare and selfishness at all costs.

> Natal enjoys swarajya, but we would say that, if we were to imitate Natal, swarajya would be no better than hell. [The Natal whites] tyrannize over the Kaffirs, hound out the Indians, and in their blindness give free rein to selfishness. If, by chance, Kaffirs and Indians were to leave Natal, they would destroy themselves in civil war. (*CW* 8: 458)

The conclusion also excoriates those who put their faith in bombs as the route to independence and hunger after industrial development in the Western model. Setting out the substance of *Hind Swaraj*, Gandhi defines self-rule as follows:

> Real swarajya consists in restraint. He alone is capable of this who leads a moral life, does not cheat anyone, does not forsake truth

and does his duty to his parents, his wife, his children, his servant and his neighbour. Such a man will enjoy swarajya where he may happen to live. A nation that has many such men always enjoys swarajya. (*CW* 8: 458)

In *Hind Swaraj* these ideas will be extended so that in theory virtually any person who practices self-rule could be Indian.[34] In this minimal definition, Gandhi had sloughed off caste, nation, language, whiteness, and body as preconditions for belonging to India, leaving religion, civilization, empire, and soul. The pamphlet collection that *Indian Opinion* generated provided one trajectory for these ideas to develop. Binding them together allows us to see Gandhi's ideas as shaped in the format of serial print culture, with ideas emerging provisionally and experimentally rather than retrospectively via the monumentality of the book.

5

A Gandhian Theory of Reading

The Reader as Satyagrahi

Hind Swaraj takes the form of a dialogue between an "Editor" and a "Reader." In the canonical traditions of Gandhian scholarship, the Editor is routinely identified as Gandhi himself, while the Reader is one or more of the Indian expatriate revolutionaries that he met in London (the two usual suspects are Shyamji Krishnavarma, editor of the radical *Indian Sociologist,* and full-time revolutionary, V. D. Savarkar).[1] In some cases the general addressee ("reader" with a lowercase "r") is recognized as including some additional constituencies, in Parel's words, "The Extremists and Moderates of the Indian National Congress, the Indian nation and 'the English.'"[2]

Yet if we turn to the foreword of the Gujarati edition of *Hind Swaraj,* the addressee is identified as "the reader of *Indian Opinion,*" a figure invoked eight times in a short text of some 350 words. The "Preface to the English Translation" likewise identifies this trusty subscriber as the addressee: "The dialogue, as it has been given, actually took place between several friends, mostly readers of *Indian Opinion,* and myself."[3]

Despite being so prominently invoked, the reader of *Indian Opinion* has disappeared from Gandhian scholarship.[4] Why? In part the answer is straightforward: as Gandhi and the book acquired prominence after his return to India, new editions dropped the South African prefatory materials. This elision cleared the way for an interpretation of the text as a debate between Gandhi and those advocating violence as the only viable anticolonial method. This view of the text gained considerable traction, especially since in a foreword of a new edition in 1921 Gandhi himself endorsed it (*CW* 22: 259–261). Today the reader of *Indian Opinion*, initially so central to *Hind Swaraj*, has all but vanished.

My intention here is not to supplant this mainstream interpretation of the revolutionary as addressee or to suggest it is wrong. Although as an aside, this insistence on the Reader as a revolutionary has always struck me as slightly odd. If one were to envisage *Hind Swaraj* as a play, it would be difficult to imagine a fire-breathing revolutionary like Savarkar taking the part of the Reader who is generally patient, polite, and biddable. Indeed, the idea of Savarkar sitting quietly through Gandhi's speeches could almost shift the text toward comedy and parody.

Be that as it may, this chapter reintroduces the reader of *Indian Opinion* not to unseat the existing interpretation but, rather, to recognize that both are correct. If "India" exists both territorially and deterritorially, then the text can as well be about the subcontinent as about some obscure diasporic locale. I pursue this line of argument in two parts: first I examine this figure of the reader in *Indian Opinion*, and then I turn to *Hind Swaraj*, a text that started off life in the Gujarati columns of the newspaper. In tracing this continuity, I pursue the theme of reading as an allegory of satyagraha.

Reading and Self-Rule

In February 1906, in the wake of Nazar's death, Gandhi increased his contributions to the Gujarati columns of *Indian Opinion*. In addition to the abridged translations, he penned a weekly column, "Johannesburg Letter," as well as contributed other shorter items of intelligence. Upon his return from London at the end of 1909, Gandhi's contributions fell off markedly (*CW* 10: 364, n. 2).

Most Gandhian scholars recognize the years between 1906 and 1909 as pivotal in Gandhi's career.[5] During this period his thinking took a radical turn with his vow of celibacy and his abandonment of any vestiges of belief in industrial progress, turning instead to a critique of capitalism. As the satyagraha campaign collapsed around him in response to his agreement with Jan Smuts in early 1908, Gandhi turned toward a more inward and spiritual view of satyagraha, establishing his second ashram, Tolstoy Farm, outside of Johannesburg. He also had perforce to accommodate shifts in his support base; abandoned by merchants who faced ruin as their shops were impounded, small hawkers (mostly "colonial born" and Tamil) with less stock to lose moved to the fore. As seen later in this chapter, this shift registered itself inversely in Gandhi's Gujarati columns: the more he lost Gujarati merchant support, the more fervently he imagined the ideal Gujarati reader.

One additional strand in this period of ferment was a radicalization in Gandhian methods of textual production, removing it from the realm of the market and the state. Upon his return from London in 1909, Gandhi stopped jobbing printing in order to devote more time to producing pamphlets that were explicitly excluded from copyright provisions, and hence the controls of the state.[6] He also began scaling back on advertising, and then in 1912 he dispensed with virtually all advertising except those for pamphlets and books that appeared in the book pages of the paper.

The account of reading and readers that follows forms part of these decisive years. In his Gujarati columns between 1906 and 1909, Gandhi constructed an ideal reader, a process in which reading became a way of thinking about satyagraha as a patient rule of the self. Patience demanded a turning away from industrial speed to do things at the pace of the body, reading being one of the activities that had to proceed in this way.

Hind Swaraj revisits these themes by presenting a fallible Reader who needs patient encouragement and training. The methods mastered by the ideal reader in the Gujarati columns are reprised for this frail Reader. It is as if the ideal reader of the Gujarati columns has graduated to become an Editor who can now train the next Reader in a relay race of attaining self-rule and autonomy.

"The Gujarati Reader"

On a weekly basis readers of Gandhi's Gujarati columns received fulsome and specific instructions on reading. The following passage captures well the tone and explicitness of this advice:

> The Transvaal Indians are at present carrying on a heroic struggle and this paper is engaged in furthering that struggle in every possible manner. We therefore deem it to be the duty of every Indian to read every line of it pertaining to the struggle. Whatever is read is afterwards to be acted upon, and the issue, after being read, is to be preserved and not thrown away. We recommend that certain articles and translations should be read and reread. Moreover, our cause needs to be discussed in every home in India.

Our readers can do much to bring this about. They can send the required number of copies of Indian Opinion to their friends, and advising them to read them, seek all possible help from them. The present issue includes a letter addressed by the Hamidia Islamic Society to Indian Muslims. We think it necessary that hundreds of copies of this number should be sent out to India. (CW 7: 156–157)

The advice is surprisingly wide-ranging and touches on how to read, what to do with the text, both materially and intellectually, and how to spread the ideas of the paper and the paper itself. In short, the passage imagines readership as a devoted apprenticeship to *Indian Opinion*. The reader not only acquires the newspaper but archives copies, recruits new subscribers, propagandizes, and acts as postal agent on its behalf.

These themes were reprised frequently in the Gujarati columns. Readers were routinely encouraged to "concentrate," "pay very close attention" (CW 7: 209), take "deep draughts" of the text (CW 8: 247), and "read . . . through carefully several times over" (CW 8: 36). Reading also entailed "pondering" (a favored Gandhian word) and memorization, both for the reader's own edification and to assist and instruct others. When imprisonments began, the columns urged readers to memorize sections of the paper so that when they visited satyagrahis in prison they could convey the substance of the paper to them (CW 7: 68). Such a reader would, in effect, become a walking, talking version of the newspaper.

The ideal reader never grows weary or bored. Reporting from his 1909 deputation to London, which was not going well, Gandhi writes: "I am tired of reporting every time that I have no news to give about a settlement. But that is what I must say again. I know, of course, that those who are perfect satyagrahis will not weary of reading this, for they are not concerned with whether or not a settlement is reached" (*CW* 10: 63).

The columns do allow that readers may occasionally become fed up. The solution is to persevere. At the end of the ninth and final installment of the less than riveting *Sarvodaya,* Gandhi noted:

> Though some may have been bored by it, we advise those who have read the articles once to read them again. It will be too much to expect that all the readers of Indian Opinion will ponder over them and act on them. But even if a [few] readers make a careful study of the summary and grasp the central idea, we shall deem our labour to have been amply rewarded. Even if that does not happen, the reward . . . as Ruskin says in the last chapter, consists in having done one's duty and that should satisfy one. (*CW* 8: 457)

Readers could be ignorant, but only in the right way:

> There are two kinds of readers: first, those who pretend to be asleep, that is to say, those who read not indeed to be enlightened but with malicious intent and in order to pick holes; the other kinds are those who really fail to see the point and are therefore truly asleep. This dialogue is addressed only to the second kind. We can wake up those who are asleep. As for the others who feign sleep nothing can be done. (*CW* 8: 136)

Those who do not read carefully and conscientiously earn a telling off: "Those who ask this question have not been reading Indian Opinion carefully . . . my explanation of the notice did include this point. I request the reader to read Indian Opinion henceforth with great care. It will not take many days to do so" (*CW* 7: 273).

The concentration of the ideal reader never lapses. The ideal reader perseveres, slowly gleaning and internalizing the paper bit by bit.

Such patient striving ensures that the messages become "imprinted" and "engraved on the [reader's] heart," to use the language of the newspaper (*CW* 5: 342; 7: 183). If the words are truly imprinted and engraved in this way, then reader and text are one. Such a reader will have a perfect grasp of the text and will be able to embody its meaning through his actions (the reader generally being imagined as male).

This idea of applying texts to circumstances forms a predominant theme in Gandhian ideas of reading. Yet this process does not simply entail a set of mechanical procedures between person, text, and action, as the verb "apply" might suggest. Rather, any action taken on behalf of the text would arise out of a deep internalization of the ideas in which text, person, and action become one. Or, to put it in the terminology of *Indian Opinion,* readers should be like "Thoreaus in miniature" (*CW* 7: 240).

Archiving

This close association between reader and newspaper is further cemented in the range of archiving practices that the paper recommended. In addition to enveloping themselves mentally with the journal, readers were required to surround themselves physically with the paper. This could entail making a scrapbook of selected articles (*CW* 4: 387), creating a file of one's own cuttings, binding together the pamphlets, and keeping back copies for cross-referencing (*CW* 5: 470).

Gandhi also recommended using the images printed in the paper as a form of uplifting interior decoration. In June 1911 the paper issued a portrait of William Hosken, chair of the European Sympathizers' Committee that had offered support for the satyagraha effort.

We should like our readers to have the portrait glazed and to hang it up in their rooms. We have noticed that Indians have on the walls of their rooms pictorial advertisements set in frames, which are issued by wine and tobacco merchants. At other times, we see meaningless pictures stuck on walls, and we are often judged by the things with which we surround ourselves. We earnestly hope that every Indian will have in his living-room only the portraits of those who have us in their debt or whose memory we wish to cherish, and that they will be careful about things [with] which they choose to surround themselves. (CW 11: 448)

When the paper issued a portrait of Dadabhai Naoroji, readers received instructions on how they should interact with the image on their walls. To merely glance at the portrait or admire it without serious contemplation came close to idolatry. The portrait required reverence and a daily recommitment to one's duty: "The true framing of it, in our opinion, would only be to engrave it in our hearts" (CW 7: 183).

It is as if readers had to "line" themselves so thoroughly with the newspaper, inside and out, that at every turn they would not only act as readers of *Indian Opinion* but be the paper itself.

Laboring for the Paper

Unsurprisingly, readers had to labor assiduously in the cause of the paper. They were enjoined to enlist subscribers (CW 5: 913-914; 2: 250), read the paper to those who were illiterate, and send copies to friends in India (CW 7: 156-157). In one case readers were instructed in the columns to discuss the paper with whites as a way of ameliorating their opinions. The piece "Suggestion to Our Transvaal Readers" is worth quoting in full:

It is especially necessary to get as many whites as possible to read the English report and to draw their attention to the Second Resolution in particular. If they read it in the right spirit, we are sure they will support the suggestion contained therein (that Indians voluntarily register). If this should happen, the Bill [Asiatic Registration Bill] would not come into force. We therefore suggest that each reader order as many copies of this number as possible and distribute them among the whites and request them to read it. In the belief that this hope of ours will be realised, we have printed extra copies of this number. Copies may be had from the head office or from our Johannesburg office. A four-penny stamp for each copy may be sent with the order. Merely to pass on a copy to a white without any explanation would be like throwing it away. It would also be necessary to explain which portion he should read. (CW 6: 351)

This may all sound unbearably onerous, but in the world of Gandhian journalism readers were tasked with almost the same responsibility as the managers of the paper. At the height of Gandhi's satyagraha campaign in India, the newspapers *Young India* and *Harijan* sold in the tens of thousands and started to render a profit. Some advisers urged Gandhi to lower the price of the papers. He refused, feeling that the cover price represented a form of commitment from the reader who had a responsibility and a role to play in the production of the newspaper. In Gandhi's words, readers should be "as much interested in the upkeep of the papers as the managers and the editors are."[7] This sentiment would have been familiar to readers of *Indian Opinion* who were likewise told "a newspaper does not mean only its editor and management; the vast majority of those connected with it are readers" (CW 10: 484).

Linguistic Skills

Over and above these superhuman characteristics (including the willingness to go to jail [CW 9: 293]), and in their spare time undertaking social work among "drunken Madrassis" [CW 5: 194-195]), readers had to be possessed of an unusual array of linguistic skills. In some cases they were fluent in English and could be instructed to read particular items in the English section of the paper (CW 5: 439). In other cases readers at one moment could speak only simple Gujarati (CW 10: 242) yet at another their Gujarati became sophisticated and they concerned themselves with the standing and institutions of the language (CW 8: 31, 194).

Moving beyond the case of English and Gujarati, the reader at times appeared to understand a number of different Indian languages. A Gujarati column entitled "To South African Indians" reads: "I say to all the Indians in South Africa that this struggle is not confined to the Transvaal" (CW 10: 217). In another, entitled "Last Message to South African Indians" (before going to jail), Gandhi addresses "Transvaal Indians," "Other Indians in South Africa," and "All Indians" as though all could hear him via the Gujarati column. In a column "Bad Habit," Gandhi upbraids Indians from one region of India who insult those from another: "We hope that every Indian who has this bad habit will give it up, if only because such behavior stands in the way of bringing all the Indians together'" (CW 8: 168). In these instances, it is as if "the Gujarati reader" expands to contain the entire Indian community and its languages within himself, as well as at times encompassing all Indians across the world.

A similar fault line runs through the reader's relationship to India. In some cases the reader knows very little about the subcontinent and stands in need of education. In other cases the reader is acquainted with hundreds of people in India and knows the region well (*CW* 7: 156–157).

On the linguistic front, it is important that there is no equivalent attempt in the English columns to interpolate this kind of conscientious reader. Aimed at supporters of the South African Indian struggle in England, Britain, and South Africa, the English section was more diffuse and impersonal in tone. The Gujarati reader, by contrast, was more than a distant sympathizer; he was a comrade in arms (of a peaceful kind) and a devoted and dutiful follower.

What might seem like a slightly improbable figure makes more sense placed alongside Gandhi's constant attempts to rise above the linguistic and community divisions that he faced. Reporting on a mass meeting, Gandhi notes that the speech of one speaker, Hajee Habib, was so "impassioned that even those who did not know Gujarati said they could follow its purport" (*CW* 5: 359). This sentiment matches Gandhi's views on Tulsidas as rising above translation: "His devotion to God was so profound that instead of translating, he poured forth his own heart" (*CW* 9: 202). *Hind Swaraj* itself was written because Gandhi "could not restrain" himself, as he says in the Gujarati foreword, an odd way, it must be said, to introduce a book whose theme is self-rule. But be that as it may, this sentiment points to the idea that if expressed with enough sincere and intense emotion, a text may find a way of rising above language divisions. As we have seen, another form that could buy such transcendent communication was poetry.

Prose versus Poetry

The Gujarati columns of *Indian Opinion* resounded with poetry, both inside and outside Gandhi's writings. This textual poetic presence arose from the widespread use of poetry as an important form in the satyagraha movement—recited and sung on platforms, gathered together in pamphlets, and enthusiastically proffered up for publication in newspapers either on a voluntary basis or in response to a competition.[8]

In a wonderful volume, Surendra Bhana and Neelima Shukla-Bhatt have collected and translated this material.[9] By their tally, between 1908 and 1914, seventy-five poems appeared, the over-whelming majority composed by Indians in South Africa, with the remainder by poets in India (either poems written before the satyagraha campaign or occasional poems in response to the events in the Transvaal and Natal or, in one case. a Gujarati translation of a section from Walter Scott's "Lay of the Last Minstrel").[10] Most of the poems were in Gujarati, although some were in Hindi and Urdu but in Gujarati script. Three were in English.

In these poems the figures of the satyagraha campaign like Parsee Rustomjee and Thambi Naidu are woven into a crowded world of epic and mythical figures. These include Ram, Sita, and Laxman from the Ramayana, and Harischandra, the heroic figure from the Mahabharata. They rub shoulders with legendary kings and rulers from medieval India and Muslim/Judaic figures like Yunus and Moses. Smuts and Botha (or Seth Botha in the language of the poems) compete with Ravana to be villains of the piece. These characters move through different orders of time, now inhabiting the Kaliyug, the Age of Vice in Hindu cosmology, and now being present at the "havoc of Karbala" in 680 at which Hussain was martyred.[11] Christian temporalities are invoked: one English poem is set to the tune of "Onward, Christian Soldiers" and was written at sea by John Andrew while he was being deported to India.

The poems are all epically extravagant, declamatory, and vocative. Heroes of the satyagraha and legendary figures from the present are addressed directly and summoned up in front of the audience or reader. They rise up from the page and conjure up an epic landscape into which Johannesburg and the events of satyagraha are sucked.

The poems are sites of pleasure and magical address that can speak immediately and powerfully to distant audiences. They are "a fire that blazed in the ocean," to take a line from one poem, "About Satyagraha," and the title of the Bhana and Shukla-Bhatt collection.[12] The sense of intense pleasure that this poetry could evoke is captured in another suggestion that came from a reader in response to the settlement with Smuts in January 1908 that appeared to the reader as a victory (a minority view, it must be said). To mark this "triumph," the reader had suggested printing *Indian Opinion* in golden letters, a proposal that draws the paper into the realm of manuscript and calligraphic traditions. Gandhi welcomed the idea but urged patience, saying that full victory had not yet been achieved. When that day came, "the hosts of heaven will come down to watch," and that would then be the time to print the newspaper in golden letters (*CW* 8: 123–124).

While the ideal Gujarati reader might dally in this zone of pleasure, his true home lay in the more austere tracts of expository prose through whose wastes the faithful reader had to be prepared to toil.

Reader as Satyagrahi

This mode of patient reading occupies a similar position to that of satyagraha in Gandhi's equations of nationhood and self-rule. The two resemble each other in many respects. Like the development of satyagraha, or "Truth," reading is painstaking, patient, and inward. Indeed, reading well fits Uday Mehta's description of Gandhian patience as "a psychological adhesive that embedded values into the self" and cultivated an expanded inner domain.[13]

In part this resemblance appears to draw on a Ruskinian view of reading as a form of work, self-discipline, and the cultivation of an inner nobility.[14] Set out most clearly in his lecture "Of King's Treasuries" in *Sesame and Lilies* (which *Indian Opinion* began advertising from 1913), the lecture seeks to cultivate a mode of reading by which to unearth "treasures hidden in books."[15] This method of reading requires humility, sincerity, and, above all, scrupulous attention, dedication, and pondering: "You must get into the habit of looking intensely at words, and assuring yourself of their meaning, syllable by syllable—nay, letter by letter." Reading is like mining, seeking gold in "little fissures."

> When you come to a good book, you must ask yourself, "Am I inclined to work as an Australian miner would? Are my pickaxes and shovel in good order, and am I in good trim myself, my sleeves well up to the elbow, and my breath good, and my temper?" . . . [the author's] words are as [a] rock which you have to crush and smelt in order to get at it. And your pickaxes are your own care, wit, and learning; your smelting furnace is your own thoughtful soul.[16]

The amount one reads is less important than the degree of concentration: "You might read all the books in the British Museum (if you could live long enough), and remain an utterly 'illiterate,' uneducated person; but that if you read ten pages of a good book, letter by letter—that is to say, with real accuracy—you are for evermore in some measure an educated person."[17]

Insight requires extended thought and contemplation: "True knowledge is disciplined and tested knowledge—not the first thought that comes."[18] Such "well-directed moral training and well-chosen reading" lead to a strengthened inner moral state, which will place the reader in a position of ethical leadership.[19]

According to Pandiri, Gandhi read *Sesame and Lilies* in South Africa, although exactly when is not clear.[20] The comparison between the two men's views on reading is nonetheless worth making, if only to highlight those aspects of Gandhi's thinking about reading that are not Ruskinian. One of these is a greater preoccupation with the theme of speed: Ruskin certainly takes the "busyness" of life and a surfeit of printed matter ("shallow, blotching, blundering, infectious 'information'") as a context but is less concerned with the tyranny of the speed at which this circulates.[21] For Gandhi, the equation of speed and efficiency lies at the heart of modernity's ills, as *Hind Swaraj* demonstrates. A large part of this thinking hence focused on slowness as a philosophy and on ways of doing things at the rhythm of the body. As an embodied experience, reading can only proceed at the pace of the body, the ideal circumstances for realizing self-rule, as *Hind Swaraj* explains. Reading might of course be perverted if done for reasons of hasty abstraction. The Gujarati reader is, however, prevented from floating off in this way by the weight of the back copies and scrapbooks he has collected. The reader's labor in archiving and being an unpaid agent of *Indian Opinion* ties reading to duty.

Like satyagraha, reading too is a radically portable skill and can best be carried out by one person at a time. As Gandhi noted in his conclusion to *Sarvodaya,* even if only one person understood the text, that would be sufficient. Indeed, even if no one read it, Gandhi would have done his duty, and the text would wait patiently until the next true reader came along. Gandhian texts and readers are never in a hurry. They do not read for the sense of an urgent, short-term ending. Writing to subscribers of *Indian Opinion* from London in 1909, Gandhi noted that even if his negotiations produced no conclusive outcome, faithful readers should persevere with his columns. The process of reading mattered more than the desire for a cathartic resolution.

Foreword to *Hind Swaraj*

This centrality of the Gujarati reader is underlined in the foreword to *Hind Swaraj.* Since this document will be discussed in some detail, I produce it here in full:

> I have written some chapters on the subject of Indian Home Rule which I venture to place before the readers of Indian Opinion. I have written because I could not restrain myself. I have read much, I have pondered much, during the stay, for four months in London, of the Transvaal Indian deputation. I discussed things with as many of my countrymen as I could. I met, too, as many Englishmen as it was possible for me to meet. I consider it my duty now to place before the readers of Indian Opinion the conclusions, which appear to me to be final. The Gujarati subscribers of Indian Opinion number about 800. I am aware that, for every subscriber, there are at least ten persons who read the

paper with zest. Those who cannot read Gujarati have the paper read out to them. Such persons have often questioned me about the condition of India. Similar questions were addressed to me in London. I felt, therefore, that it might not be improper for me to ventilate publicly the views expressed by me in private.

These views are mine, and yet not mine. They are mine because I hope to act according to them. They are almost a part of my being. But, yet, they are not mine, because I lay no claim to originality. They have been formed after reading several books. That which I dimly felt received support from these books.

The views I venture to place before the reader are, needless to say, held by many Indians not touched by what is known as civilization, but I ask the reader to believe me when I tell him that they are also held by thousands of Europeans. Those who wish to dive deep, and have time, may read certain books themselves. If time permits me, I hope to translate portions of such books for the benefit of the readers of Indian Opinion.

If the readers of Indian Opinion and others who may see the following chapters will pass their criticism on to me, I shall feel obliged to them.

The only motive is to serve my country, to find out the Truth, and to follow it. If, therefore, my views are proved to be wrong, I shall have no hesitation in rejecting them. If they are proved to be right, I would naturally wish, for the sake of the motherland, that others should adopt them.

To make it easy reading, the chapters are written in the form of a dialogue between the reader and the editor.[22]

The foreword characterizes the Gujarati readership of *Indian Opinion* as a series of concentric circles. The core comprises eight hundred subscribers surrounded by those who do not subscribe but borrow the paper and read it "with zest."[23] Together these two groups number about eight thousand: ten casual readers for every dedicated subscriber. Beyond these two circles are "those who cannot read Gujarati [and] have the paper read to them": it is unclear whether this sentence refers to illiterate Gujaratis, or to those who can read other languages (English, Tamil, Hindi) but not Gujarati, or to both.

The concentric circles of this group of Gujarati and/or other readers eddy further outward, since communities in London, Europe, and India share their views. Indeed, the text seems to imply that there may be some equivalence between illiterate Gujaratis in South Africa and the thousands in India untouched by civilization whose views the book purports to articulate.

This Gujarati group and their proxies help produce the text in key ways. Their questioning has sparked the text in the first place and has assisted in the process of bringing Gandhi's views to light: the English edition represents the book as a transcription of conversations between Gandhi and his subscribers. Their text-enabling capacities are further apparent in the fact that Gandhi intends undertaking, on their behalf, translations of the key texts used in formulating the argument of *Hind Swaraj*. It seems appropriate, therefore, that it is to this group that the text is offered by being "placed before them" (in the language of the foreword).

The Gujarati reader (having done an apprenticeship in the columns of *Indian Opinion*) understands well the Gandhian modes of text-making outlined in this foreword, notably, the careful reading and pondering of a series of books that becomes internalized ("They are almost a part of my being").[24] The Gujarati readers who have time (and which Gujarati reader skilled in patient reading would not?) can reproduce this process by "diving deep" and reading these same books for themselves.

This intimate relationship of text-maker and reader is mirrored in the structure of the foreword itself, which proceeds in a sandwich-like way, alternately layering reader and text-maker. The opening sentence addresses the reader of *Indian Opinion* before describing how *Hind Swaraj* itself was made. We then proceed to a description of the Gujarati subscribers and readers before returning to the making of the texts by way of some comments on the books that have informed Gandhi's ideas in *Hind Swaraj*. The discussion then interlards another "layer" of readers, this time the many Indians untouched by civilization, and the Europeans who share their views, before returning again to the list of books that informed *Hind Swaraj*, which appears at the end of the text. Reader and text-maker alternate in a generative sequence in which the one enables the other.

Importantly, the kind of text-making described in the foreword constitutes less a form of writing than editing. The foreword flags that this book "lays no claim to originality": it is not the expression of some individual genius.[25] Instead it has evolved through the tailoring and retailoring of texts: reading, pondering, internalizing, discussing, and translating. We can hence read the layered sequencing of the foreword as moving between Reader and Editor. Having demonstrated this alternation in action in miniature, the foreword concludes by announcing this as the method of *Hind Swaraj* itself: "To make it easy reading, the chapters are written in the form of a dialogue between the reader and the editor."[26]

Tableau of Reading

Hind Swaraj presents a tableau of reading that is situated in no specified place. Our work as readers is to imagine where it might unfold. One way to answer this puzzle is to look carefully at the opening exchange between Reader and Editor. In this passage, the Reader asks a straightforward question:

> Just at present there is a Home Rule wave passing over India. All our countrymen appear to be pining for National Independence. A similar spirit pervades them even in South Africa. Indians seem to be eager after acquiring rights. Will you explain your views in this matter?[27]

In what will become a characteristic move, the Editor sidesteps the question. Instead of offering a direct response, he explains how he will answer. His method will be that of the newspaper:

> You have well put the question, but the answer is not easy. One of the objects of a newspaper is to understand the popular feeling and to give expression to it: another is to arouse among the people certain desirable sentiments; and the third is fearlessly to expose popular defects. The exercise of all these three functions is involved in answering your question. To a certain extent, the people's will has to be expressed; certain sentiments will need to be fostered, and defects will have to be brought to light. But, as you have asked the question, it is my duty to answer it.[28]

The dialogue is hence less a conversation between Reader and Editor as real people than it is a textual encounter in a Newspaper in which the Editor can only speak through the genres and forms of that medium. The encounter is something of a contest in which the Newspaper, through its textual strategies, must tame an initially resistant Reader, changing him or her from a hasty and impatient interlocutor to a patient and considered individual. *Hind Swaraj* constitutes a dramatized, talking advice manual that seeks to inoculate readers against dangerous reading situations.

Two such situations faced the readers of *Indian Opinion:* the first entailed readers being confronted with wrongheaded and dangerous intelligence, the second with too much information that might overwhelm them, seducing them into hasty and ill-considered reading. With regard to the first danger, the periodicals of the revolutionary extremists, like *Bande Mataram* and *Indian Sociologist,* had started to make their way to South Africa and to "infect" some sections of the Indian community. On the second count, what *Hind Swaraj* terms a "flood of Western thought" in the form of books and newspapers plagued many societies and was beginning to bear down on the Indian community.[29] In the view of *Hind Swaraj,* there are too many newspapers, expressing too many fickle and contradictory ideas. There is a surfeit of information and ideas where anyone can make their views known. "Formerly the fewest men wrote books that were more valuable. Now anybody writes and prints anything he likes and poisons their minds."[30] Just as the unfettered flow of trains brings disease and famine, so anyone can "abuse his fellows by means of a letter for one penny."[31] Unless one is inoculated against it, print culture can suck one into the addictive consumerism of industrial capitalism and the pathologies of those "infected with the machinery craze."[32]

The Reader in *Hind Swaraj* stands midway between these two positions, harboring some half-digested ideas of the revolutionaries and attracted to the trappings of English "civilization." The Editor has to wean him from both of these ill-considered positions by showing him how to read carefully, thoroughly, patiently, and for himself.

Reprising *Indian Opinion*

Hind Swaraj arises from the textual environment of *Indian Opinion*. It reprises the forms from which it comes and assists the Reader's education through making him repeat some of the pedagogic techniques refined in the columns of the paper.

Q and A

The dialogue form of *Hind Swaraj* rises up from the pages of *Indian Opinion* that are replete with dramatized and direct speech. Examples include the interrogative headline, a widespread, almost-signature feature of the paper: "What Is Habitation?"; "Are Indian Peoples British Subjects?"; "What Is Hawking?"; "Are British Subjects Serfs?"; "What Is a Hawker?"; "Is a Verandah a Shop?"; "What Is a Coloured Person?"[33] Reports on trials, speeches, and mass meetings are likewise set out in reported speech.[34] Advice on how to deal with the challenges of satyagraha appeared in a Q and A format: what to do when one received letters of notification, how to act during a police investigation, and so on (CW 6: 459–462; 7: 6).

In other cases the newspaper used Gujarati traditions of the dialogue as a literary form. In 1911, from July 29 to December 23, the Gujarati section of the newspaper serialized a twenty-part story entitled "Introducing South Africa or the Dialogue of Two Friends, by an Indian." The piece is a conversation between Uday-shanker, an old South African hand, and his childhood friend, Manharram, who has just arrived in Natal, giving advice on life in

Durban, how to find a job, where to shop, how to write letters, and matters of social etiquette.[35]

Early in 1908, Gandhi extended these techniques in a sustained dialogue between a "Reader" and an "Editor" that rehearsed the form he would use in *Hind Swaraj* (*CW* 8: 136–147). Entitled "The Dialogue on the Compromise," the piece sought to address the confusion and anger that Gandhi's sudden agreement with Smuts had precipitated.

In this muddled atmosphere, Gandhi had been confronted with questions, letters, and accusations, a situation that prompted him to select dramatized dialogue: "We therefore answer most of the questions [sent to us] in the form of a dialogue" (*CW* 8: 36). The piece is hence meant to function as a distancing and depersonalizing aggregate of views. The Editor importantly is not Gandhi, since the two interlocutors discuss "Mr Gandhi" as a third person and they likewise talk about *Indian Opinion* as though they are not in it. The dialogue consequently asks its readers to take a longer and more distant perspective on matters. The Editor also insists that the Reader must bring good faith to the interaction, asking questions "in the presence of God, with sincere and patriotic intent" (*CW* 8: 137).

While such admonitions may today sound patronizing and hectoring, they form part of the expectation of the ideal reader as possessing self-control and persistence (the dialogue, needless to say, has to be read through "carefully several times over" [*CW* 8: 136]). The Reader needs ethical character and substance, both as part of his journey of self-development but also to enable a reasonable exchange of views. As scholars of Plato's dialogues have indicated, for the shared and "maieutic" ("midwifely") pursuit of dialogue to proceed in a Socratic manner, both participants need a similar ethical orientation.[36]

Gandhi's dialogue, *Hind Swaraj,* redacts these experimental forms from *Indian Opinion.* In the Gujarati preface, Gandhi explains that he has used the dialogue format for "easy reading," a claim that reprises both his use of Q and A to summarize information and his earlier dialogue on the compromise to provide distance and clarity on pressing issues. In the English preface, he indicates that he chose the dialogue form, since "the Gujarati language readily lends itself to such treatment, and that it is considered the best method of treating difficult subjects."[37] This claim looks back both to Gujarati literary tradition and *Indian Opinion*'s use of it in "Introducing South Africa or the Dialogue of Two Friends, by an Indian" and "The Dialogue on the Compromise," as well as the Q and A forms surrounding them.

In the task of dramatizing the education of a Reader, *Hind Swaraj* reprises a further range of formats from *Indian Opinion.*

Binding Biographies

As part of the opening discussion in *Hind Swaraj,* the Editor points to the Indian National Congress as the progenitor of home rule. Sharing the view of young revolutionaries, the Reader impatiently brushes aside Congress and all of its doings. In response the Editor compiles a gallery of prominent Congress figures (Naoroji, Hume, Wedderburn, Gokhale, Tyebji), rather like a miniature version of the biography series that ran in *Indian Opinion.*[38] As we saw, this series was accompanied by advice to readers to compile a scrapbook so that they could read and reread about these exemplary lives.

The dialogue reenacts this process. The Reader is unimpressed by the notables on the list and pooh-poohs them as collaborators of no consequence. The Editor, however, brings the Reader back to

the Congress figures, this time expanding in detail upon the achieve-
ments of each (or at least the first three).[39] In directing the Reader's
attention to material already covered, the Editor demonstrates the
importance of patient consideration, of returning repeatedly to a
text until one has understood it.

This repetition starts to take effect and the Reader "begins to understand somewhat" and by the opening of chapter 2 has accepted
that "the foundation of Home Rule was laid by the Congress."[40] A
discussion of the partition of Bengal follows, and by chapter 4 the
Reader proves he has been listening carefully by rehearsing a summary of the debate thus far: "I have now learnt what the Congress
has done to make India one nation, how the Partition has caused
an awakening, and how discontent and unrest have spread through
the land."[41]

This strategy of compilation recurs elsewhere. In trying to "un-
deceive" the Reader about the true nature of industrial civilization,
the Editor provides a series of graphic examples demonstrating
how bodily welfare has become the object of life.[42] People occupy
ever-grander houses, acquire more and more clothing, plow with
steam-powered engines, fly about in trains, kill with machine guns,
and labor like slaves in factories. The rueful refrain "This is civilization" punctuates the litany of examples.[43]

This repetition not only impresses the reader who promises to
"ponder" the issue, it also establishes an important principle in
the text, namely, that understanding needs to be rooted in concrete
example or, better still, beaten out on the anvil of experience. The
text is skeptical of grand abstract schemes in which freedom can
be claimed on behalf of others. "What others get for me is not Home
Rule" as the text famously says; people have to claim their own
freedom by learning to rule themselves.[44] In the same way, one has

to learn to read for oneself through careful, painstaking work. Glibly spouting the ideas of others not only wastes one's time but can be dangerous. In Gandhi's estimate, Madan Lal Dhingra, who assassinated Sir William Curzon-Wylie on July 1, 1909, in part acted on the basis of "ill-digested reading of worthless writing" (CW 9: 428).

The appeal of these dangerous ideas is, however, demonstrated in the text by the Reader who backslides from time to time under their sway, becoming impatient, both in his dealings with the Editor and in his desire to see the English leave India. "In your excitement, you have forgotten all we have been considering."[45] The parabola of the Reader rises and falls markedly. In the first quarter of the text, the Reader is a bit bolshie; in the second, he calms down, but in the discussion of Italy and India he again goes into revolt, before being calmed by the discussion of passive resistance, education, and machinery. The Editor repeatedly urges patience both politically and in terms of reading. Hasty demands and schemes will have no lasting value, just as thoughtless and ill-digested reading is counterproductive.

The Reader starts to glimpse the benefits of repetition and agrees that understanding takes time and "digestion": "You have set me thinking, you do not expect me to accept at once all you say. You give me entirely novel views. I shall have to digest them."[46] The Editor is aware of the limits of his power: he can place ideas before the Reader and draw his attention to them; he can offer lessons in repetition and reading, but the Reader has to read for himself.

The process of understanding true swaraj, or self-rule, is equally the process of learning to read in a patient, concrete, nonteleological way.[47] As Mehta argues, the dialogue "is at pains not to allow [Gandhi's] disagreements with his interlocutor to assume an abstract form, as though what were at stake were simply two different positions, disembodied from their advocates." Instead, Gandhi "urges patience in an almost intimate voice, suggesting that what he has in mind has to be felt and not merely read."[48] Such deep understanding has to arise both in and beyond the text, in some sense appropriately mirroring debates in Plato's *Phaedrus.*[49] Here, "ensouled" knowledge lies beyond writing, a medium with limits, which can create a "spurious appearance of intelligence," and famously is unable to give an explanation of itself, "just go[ing] on and on for ever giving the same single piece of information."[50] As Christopher Gill points out, the Platonic dialogue uses writing in a way that offsets these drawbacks by making conscientious readers think for themselves and enabling them to take the argument forward beyond the written dialogue itself.[51]

All the Editor can do in the end is "place [the text] before the reader," a phrase that recurs frequently in *Hind Swaraj* and elsewhere in Gandhi's writing. Deceptive in its simplicity, the term distills much Gandhian thinking on reading as a paradigm of ideal social relations. The act of placing something before a reader carries overtones of a gift. Such gift-giving creates relationships of obligation between the Reader and Editor and shifts reading away from the realm of anonymous mass consumerism and toward the domain of semipersonalized reception.

Placing an object before a group equally carries a certain ceremonial association, which turns a motley and dispersed audience into a purposive textual congregation. In this commonwealth, all can find a place provided if of course they are prepared to undertake the necessary apprenticeship of reading. In a Gandhian theory of reading, those who do so with virtue and application will turn themselves into true readers and writers, exemplars and analogues of self-ruling subjects, and miniature and summarized zones of sovereignty.

Gandhi's experiments with the Gujarati reader form part of the ferment and flux of his South African years. Once in India, and as anti-imperialism came to assume the form of anticolonial nationalism, he had to abandon the more radical aspects of his open-ended and deterritorialized ideas of self-rule. This process began as he left South Africa. On board the SS *Kinfauns Castle*, he wrote a retrospective on the satyagraha campaign that was destined to appear in the Gujarati columns and then in the *Golden Number of Indian Opinion: Souvenir of the Passive Resistance Movement in South Africa 1906–1914* (whose cover, incidentally, appeared in gold lettering). The opening sentences read: "During the last campaign, the very highest limit was reached. I have had simply no time to write of the experience. I had meant to share it with the readers of Indian Opinion" (*CW* 14: 265).

These lines invoke the previous occasion on which Gandhi had addressed the Gujarati reader at sea, namely, during his writing of *Hind Swaraj* in November 1909, on board the SS *Kildonan Castle*. In this case the figure of the Gujarati reader sat at the center of his attention and called the book into being. Five years later, in 1914, the Gujarati reader is starting to recede, and Gandhi is already forgetting to share things with him. As he headed to London and then India and to a more nationalist future, the figure he had invented to elude a narrow idea of nation remained behind. Embedded in the columns of the newspaper, "the Gujarati reader" still waits patiently for a reader who can bring him to life.

Conclusion

"No Rights Reserved"

Today it is virtually impossible to locate a copy of the first edition of the English version of *Hind Swaraj* published at the International Printing Press (IPP) under the title *Indian Home Rule*. The Johannesburg Public Library once owned a copy but it has disappeared, and none of the world's major deposit libraries has any in their holdings. One survives in South Africa among Gandhi's descendants, but no public copy is available either in the country or beyond.

In part, this disappearance is not surprising: like all printed ephemera, pamphlets seldom endure. Yet there is something larger in this paradox that the first edition of one of the modern world's most important books is absent from the world's major libraries. As Gandhi's key manifesto, the text itself has become legendary and exists in dozens of formats. However, this most important of editions barely survives.

Its tenuous existence is well illustrated by a visit to WorldCat. A search under both titles (*Indian Home Rule* and *Hind Swaraj*) produces a list of fifty-three "editions and formats," commencing with the first Indian edition of the book produced in Madras in 1919. Seen from this perspective, *Hind Swaraj* only starts to exist when it assumes the form of a "book" in India, a space defined by a territorial state. But if one repeats the search adding the key word "Phoenix," an entry for the 1910 first English edition pops up. This item promises that there are libraries in Switzerland, the Netherlands, and the United Kingdom with copies. If one pursues these links, one will find that these libraries hold editions only from the 1920s.

The first English edition is both there and not there, momentarily becoming visible before disappearing again. This spectral quality is appropriate for a pamphlet wrought in the context of Gandhi's heterogeneous textual methods, which took shape between empire and nation; the deterritorial and the territorial; the periodical/pamphlet and the book; free circulation and copyright; and South Africa and India. Contemporary editions of *Hind Swaraj* in India still carry signs of this heterogeneity. While bearing some of the appurtenances of bookness, namely ISBN numbers and copyright statements, these editions have the artisanal feel and price of pamphlets, with some, like those published by Navajivan, being subsidized.

The mixed character of these contemporary editions usefully points to the trajectory of Gandhi's textual work on his return to India. On the one hand, he deepened his South African periodical experiments in his papers *Young India, Navajivan,* and *Harijan* while, on the other, he formalized his writing through book production and a limited engagement with copyright. In terms of his periodical production, aspects of his South African experience were developed and intensified: advertisements fell virtually entirely by the wayside, while Gandhi's publications became even less like newspapers and more like ethical anthologies—"viewspapers," as he famously described them.[1] His papers refined themselves as vehicles for propagating the "inner meaning" of satyagraha, as well as being one of its methods.[2] The duties of the reader remained onerous. As indicated earlier, when the circulation of *Harijan* and *Young India* increased, some advisers suggested dropping the price. Gandhi demurred: the reader had as much responsibility as the editorial team or management and so should continue to make a greater rather than lesser financial contribution. As in South Africa, the reader had to be prepared to embody and even *be* the paper. Facing the prospect of closure because of colonial state pressure, *Harijan* urged readers to become "walking newspapers," passing the "news" of satyagraha from person to person.[3]

Yet alongside this periodical production, Gandhi continued to be "booked." From the 1920s, anthologists within India and further afield started to harvest collections from the vast textual fields that Gandhi had created.[4] At the same time, aspects of his own work migrated into hardback form in the subcontinent and beyond, as *Hind Swaraj, Satyagraha in South Africa,* and especially his autobiography were produced as copyrighted objects for an international market.

Through Gandhi's long career as a publicist, his writings have accumulated across the world in textual fields that are as extensive as they are heteromorphous. Yet the first English edition of *Hind Swaraj* has escaped even this capacious and sprawling "archive." Conceptualized as an experimental text outside the grid of the nation-state and copyright, the pamphlet has eluded the normal methods of "capture" devised by such institutions as deposit libraries.

The precise character of the text's experimentation emerges from two features in the first edition, namely the phrase "No Rights

Reserved" on the title page and the invocation of the reader of *Indian Opinion* in its preface.[5] The latter invokes the world of the diasporic subject, the former the realm of free circulation.

Taken together, these features denote that the reader of *Indian Opinion* is being advised that there are "No Rights Reserved," a phrase that not only announces a refusal to be part of a copyright order but also indicates that anyone has the right to reproduce the text.[6] In doing so, the reader of *Indian Opinion* would quite literally be reproducing "Indian Home Rule." Put differently, the lowly diasporic subject would be replicating Indian sovereignty. This reproduction happens both in the reading of the pamphlet (provided of course that the reader follows the instructions in the text) and, as the reader becomes a satyagrahi, in the ability to teach the lessons of the book to others. As the reader graduates to become an editor, he or she could in turn physically reproduce sections of the book to disseminate it as widely as possible.

This material practice of reproducing texts furnishes yet another instance of Gandhi's theory of sovereignty rooted in the individual. South Africa furnished an important context for the shaping of these ideas. Here Gandhi encountered the lowly diasporic subject and had to compute how he or she could be a potential locus for the nation. The phrase "No Rights Reserved" and its different meanings usefully contain the conclusions at which Gandhi arrived. On the most obvious level, the phrase indicates that in Gandhi's vision of India there could be no talk of allocating rights exclusively to some groups and not others. The ambit of the nation should be as wide as possible, taking in even the lowly diasporic subject.

One older meaning of "reserve" derives from the world of the church and the law and entails assigning a case or judgment to be dealt with by a superior authority. In such cases, decisions are reserved to the bishop, pope, senior judge, and so on. Here rights are conferred on an elite who can in turn decide to delegate them lower down. In a Gandhian world, such assigned rights were of course not worth having. In the famous formulation in *Hind Swaraj*: "What others get for me is not Home Rule." Rights reserved to one group and awarded to another are no rights at all.

Copyright subsists in the right to copy, an entitlement conferred by the market and the state. The "No" in the phrase rejects the authority of these two institutions and encourages the reader to exercise this right to copy on his or her own behalf. This shift removes readers from the abstract realm of the market and the state and involves them in a concrete activity of making copies, whether physically or mentally. Rather than the state holding copyright over sovereignty, which it may or may not confer on others, subjects have to seize this right for themselves by making it tangible.

A deterritorialized diasporic subject (the reader of *Indian Opinion*) out on the peripheries can make himself or herself a sovereign part of India not through territorial belonging or abstract rights but through reading (and reproducing that reading) in a system of free circulation. Needless to say, as anti-imperialism and hence nationalism gained ground, this coupling of the diasporic subject and unrestricted circulation as a sovereign expression of India had little chance of survival.

While India had been imagined as a dominion in empire, diasporic communities could acquire symbolic capital (as the South African Indian struggle did). However, as India became more national, the diasporic fringe lost much of its value and at times turned into something of a political liability. Diasporic subjects continued to articulate ideas of Indians as more worthy claimants of imperial rights than other groups like Africans. At a time when ideals of anti-imperial solidarity between different colonized groups became orthodox, these older claims based on ideas of civilization became an embarrassment to anticolonial leaders in India.[7] Under such circumstances, the idea that lowly diasporic subjects may have played a role in the formation of India became unthinkable in the same way that the reader of *Indian Opinion* might be the figure calling Gandhi's key text *Hind Swaraj* into being has been discounted. Furthermore, the notion that such a lowly subject in some distant part of the world was entitled to a system of free circulation in which he or she could exercise the right to copy gained no traction.

However, during his South African years, Gandhi seriously contemplated precisely these conjunctures. As a deterritorial unit, India could best be defined by free textual circulation rather than territory. Within this zone, sovereignty could occur through individuals who turned themselves into satyagrahis via practices like reading or textual production. But as nationalism became an orthodoxy, these utopian imaginings have receded almost totally from view. As we enter an era more sympathetic to transnational forms of analysis, Gandhi's distant utopian horizons once again become visible. Working in this climate, this book has sought to secure Gandhi's South African textual experiments a safer passage and a better hearing.

By recognizing speed, summary, and discontinuity as ineluctable conditions of modern reading, Gandhian theories of text have much to say to our current conditions where the tempos of industrialized information have sped up beyond any imagining. As the reverberations of this acceleration judder around us, one site of "crisis" has been around the future of the "book," or novel, and the continuous reading with which it is so strongly associated. There is a mounting concern that we are entering (or indeed reentering) an age of Internet magpie reading, a bit of this and a bit of that. Shadowing these apprehensions come larger anxieties about copyright, intellectual property, authorship, and plagiarism.

The attention lavished on David Shields's *Reality Hunger* illustrates the range and depth of these apprehensions.[8] The book itself celebrates the end of the novel in favor of more "real" forms like the lyric essay, memoir, and autobiography, accompanied by less reverence for originality and intellectual property. The form of the book itself is a series of short paragraphs, mostly unattributed quotations, resembling nothing so much as a snippets section from an Edwardian periodical. As Luc Sante points out in his review of the book, Shields is a "wisdom-junkie" who wants "a literature built entirely out of contemplation and revelation" and thinks that "Hamlet would be a lot better if all the plot were excised, leaving the chain of little essays it really wants to be."[9]

Reality Hunger takes us back to the world of the periodical and anthology, out of and in opposition to which structures of authorship, intellectual property, and originality arose. In part, Shields's anthology can resonate with these anxieties since, like many grand structures of the "West," they are collapsing under their own weight. Too expensive to maintain, with too little purchase in a non–Euro-American world, and requiring too much surveillance, these underpinnings of a historically specific textual economy are unsustainable. Everything appears to be sinking back into the lower empire of periodicals where—if we turn to Gandhi—it has always been. This world of recycled text, impersonal writing, and reader-as-redactor has typified the experience of most "ordinary" print culture users.[10] This is the textually demotic condition that has obtained throughout most of the world, especially in the third world or the global south.

This return represents a textual leveling of the world, with the north coming to resemble the textual ways of the global south, where such forms of textual production, notably the periodical and the pamphlet, have always been the norm. Unable to sustain the resource and surveillance-intensive intellectual property regimes and ideals of authorship that these underwrite, the north faces a textual crisis that mirrors its debt catastrophe. As Jean and John Comaroff argue, the north is indeed evolving southward.[11]

Achal Prabhala, a scholar of international intellectual property, has recently made the same point in arguing that Gandhi would have been a Wikipedian. The model of open-ended, cooperative knowledge enabled by the various Wikimedia projects demon-strates "that knowledge does not have to be a passive act, and that the value of generosity can be productive and revolutionary at once."[12]

Gandhi, he argues, was a similar kind of "free knowledge activ-ist." The evidence that he uses is a photograph of the first English edition of *Hind Swaraj* in the possession of Uma Dhupelia-Mesthrie and photographed by me. As an international expert on copyright, Prabhala fully grasped the significance of the phrase "No Rights Reserved," which appeared on that title page.[13]

A Gandhian theory of text illuminates these changing textual economies by making it evident that discontinuous reading has and always will be the most widespread form of reading. A Gandhian reality focuses more on how, rather than on what, one reads. Or, as Ruskin argued, it is better to read ten pages carefully than to rush thoughtlessly through all the books in the British Museum. This is not simply a matter of intensive versus extensive reading—Gandhi was clear that one had to do both. Rather, what becomes cardinal is slowness, not for its own sake but as an antidote to the automatic equation of speed with efficiency. One cannot outsource serious reading or pretend to read more or faster than one's body permits one. Such Gandhian ideas of slow reading could usefully be absorbed into the idea of slowness as a politics that is emerging across a number of domains: slow food, slow cities, slow schooling, slow sex.[14]

There are of course very real limits to Gandhian theories of reading, which are powerfully bound up with ideas of interiority. Postmodernity now operates as modernity without interiority, a series of instantaneously assimilable, visual signs and experiences. From a Gandhian perspective, such turbo-reading would represent "macadamization" taken to absurd lengths: to those living in the swirl and confetti of social media, he would no doubt quote Thoreau, namely, that they "have not heard from [themselves]" in a long time.[15]

Returning to *Reality Hunger*, there is a second dimension that is relevant to this discussion, namely, the return to didacticism. As Sante's earlier comments indicate, Shields takes us back to the world of the didactic nugget, a form that, as Leah Price argues, has long received bad press as a result of the incremental feminization of didacticism. One agent in this process has been the anthology, which sifted, valued, and gendered narrative and nonnarrative material in changing ways, men initially being associated with philosophical nuggets, women with frivolous narrative, and then a reverse as women readers were lumped with uplifting maxims, men with "efficient rush through plot." This literary triage finally resulted in bad press for "feminine moralism," contrasted unfavorably with the "narrative pleasure of masculine romance" à la Stevenson or Kipling.[16]

For long the poor cousin in the academy, the abject opposite of high modernism and literary complexity, didacticism seems set to make a comeback, at least if Shields is to be believed. As an ardent ethical anthologist, Gandhi intersects with this story. Placed back into a realm where didacticism is no longer cold-shouldered, Gandhi's English work may be able to receive a more considered literary interpretation. While his Gujarati writings have received fulsome coverage (Gandhi's name graces a period in Gujarati literary history), his English works, with the exception of his autobiography, have surprisingly seldom been taken as literary objects in their own right.[17] In part, this has to do with the perception that Gandhi's writings are journalistic and hence not sufficiently literary to merit close scrutiny. A second factor has been the hagiographical aura around the Mahatma in which his words are seen as secondary to his life or his inner spiritual state so that his writing can only ever have a derivative, reflective status.[18] Where the phrase "Gandhian literature" is used, it represents discussions of Gandhi's influence on others or his representation in fiction or other media.[19]

This book has attempted to demonstrate the richness that lies in approaching Gandhian texts via a literary route. In some senses the style of reading we call literary is somewhat Gandhian: slow, scrupulous, patient, repetitive, cumulative.

Applying such approaches produces a useful alignment where text and method of reading become companions, assisting each other along the way. In this book, one destination of that journey has been a deeper appreciation of Gandhian theories of reading and their radical import. In a Gandhian world, serious reading, both in its content and method, can have far-reaching implications: slowing down the headlong rush of the industrial machine to the pace of the human body, pausing the addictive tempos of market-driven life, and creating small zones of independence outside the realm of the nation-state. In these equations, reading sits at the heart of a new theory of sovereignty. The claims that Gandhi makes on behalf of reading are ambitious and have implications for a range of disciplinary areas. To use a Gandhian phrase, this book places these ideas before the reader, hoping that he or she will recognize the potential of reading as a central heuristic category and take it forward—slowly and carefully of course.

APPENDIX

Pamphlets Reprinted from *Indian Opinion*

English

J. L. P. Erasmus. *The Story of the Ramayana*. Phoenix: International Printing Press, 1905.

For Passive Resisters. Phoenix: International Printing Press, 1907.

M. S. Maurice. *Ethics of Passive Resistance*. Phoenix: International Printing Press, 1907.

[H. S. L. Polak]. *A Book and Its Misnomer*. Phoenix: International Printing Press, 1907.

M. K. Gandhi. *Indian Home Rule*. Phoenix: International Printing Press, 1910.

Leo Tolstoy. *Letter to a Hindoo*. Phoenix: International Printing Press, 1910.

Gujarati Pamphlets

As this list is drawn from reports on pamphlets (advertisements, reports on banning, etc.) rather than the booklets themselves, the dates of publication are not known.

BY GANDHI

Hind swaraj.

Maro jail no anubhav (My Jail Experience). Published in three parts.

TRANSLATIONS BY GANDHI

Ek Hindu pratye Tolstoy no kagal (Tolstoy essay, *A Letter to a Hindoo*).

Ek satyaveer ni katha (The Story of a Soldier of Truth; Plato's *Apology*).

Jeevan dori (Thread of Life; translation of Tolstoy, possibly *On Life*).

Manus ketli jameen no mālik hoi shake? (How Much Land Can a Man Own as Proprietor?; Tolstoy story, *How Much Land Does a Man Need?*).

Murakhraj (The King of Fools; Tolstoy story, possibly *Ivan the Fool*).

Mustafa Kemal Pasha nu bhāsan (Speech of Kamal Pasha).

Mustafa Kemal Pasha nu jeevan charitra (Life of Kamal Pasha) (translated from clippings from Egyptian newspapers).

Niti dharm athwa dharm niti (Ethical Religion or Religious Ethics; translation of William Salter, *Ethical Religion*).

Sarvodaya (The Advancement of All; John Ruskin, *Unto This Last*).

Satyagraha (M. S. Maurice, *Ethics of Passive Resistance*).

Satya vina beeju kashu nathi (There Is Nothing but Truth; Tolstoy essay, *None Other Than Truth*).

Suhdara na bhavāda (The Failure of Reform; translation of articles of Justice Syed Ameer Ali).

BOOKLETS ON VARIOUS TRANSVAAL LAWS TRANSLATED BY GANDHI
(POSSIBLY APPEARED IN *INDIAN OPINION*)

Transvaal na nava kayda mujab na dara (Provisions According to
the New Transvaal Law).

Transvaal no dukan band karvāno kaydo (Transvaal Law Regarding the
Closing of Shops).

Transvaal no khooni kaydo (The Murderous Law of the Transvaal).

Transvaal no navo kaydo (The New Law of the Transvaal).

OTHER GUJARATI PAMPHLETS

Jail nā kavya (Jail Poems, collection comprised of entries for poetry
competition run by *Indian Opinion* [*CW* 7: 10]).

Lo. M. Bal Gangadar Tilak tém na mat ané veechar (The People's Champion
Bal Gangadar Tilak—His Opinion and Thought).

There were as well pamphlets printed by the International
Printing Press without first appearing in *Indian Opinion*. These include
Besant's *Gita, Jasmāni ni garbi* (Jasma's Poem), and the first canto of Tulsi-
das's Ramayana in Hindi. In other cases, material was compiled from the
newspaper for commemorative purposes, such as *Deshbhakta Gopal Krishna
Gokhale* ("Description of Mr Gokhale's Visit to South Africa in Which
Pictures Are Taken during His Travels, Pictures of Hindi Leaders, General
Botha, General Smuts and Testimonials Received in Honor Are Printed
on Beautiful Paper"). The English version is *Indian Opinion Souvenir of the
Hon. Gopal Krishna Gokhale's Tour in South Africa, October 22nd–November
18th*. Durban: Indian Opinion, 1912. Another pamphlet was *Deshbhakta
Gopal Krishna Gokhale ané girmit na savāl par ni charcha* (Patriot Gopal
Krishna Gokhale and the Discourse on the Indentured Question).

Notes

Introduction

1. Benedict Anderson, *Imagined Communities: Reflections on the Origin and Spread of Nationalism* (London: Verso, 2006), 86.
2. Susan Bayly, "Imagining 'Greater India': French and Indian Visions of Colonialism in the Indic Mode," *Modern Asian Studies* 38, no. 3 (2004): 703–744.
3. M. K. Gandhi, *Hind Swaraj or Indian Home Rule*, ed. Anthony J. Parel (Cambridge: Cambridge University Press, 2009), preface, 7.
4. Uday S. Mehta, "Patience, Inwardness, and Self-Knowledge in Gandhi's *Hind Swaraj*," *Public Culture* 23, no. 2 (2011): 417–429.
5. See Parel's introduction to *Hind Swaraj and Other Writings*, ed. Anthony J. Parel (Cambridge: Cambridge University Press, 2009), xliii–xliv, for a useful discussion of Gandhi's use of authors who criticized the importance of speed in nineteenth-century industrial life.
6. *Collected Works of Mahatma Gandhi.* http://www.gandhiserve.org/cwmg/cwmg.html (accessed January 2, 2006).
7. Ananda M. Pandiri, *A Comprehensive, Annotated Bibliography on Mahatma Gandhi*, vol. 1 (Westport, Conn.: Greenwood Press, 1995), 155–242.

8. Shyam Balganesh, "Gandhi, Freedom and the Dilemmas of Copyright Law," talk delivered to the Centre for Internet and Society, Bangalore, January 30, 2012.

9. On indenture, see Hugh Tinker, *A New System of Slavery: The Export of Indian Labour Overseas, 1830–1921* (London: Oxford University Press, 1974); David Northrup, *Indentured Labor in the Age of Imperialism: 1834–1922* (Cambridge: Cambridge University Press, 1995); Ashwin Desai and Goolam Vahed, *Inside Indenture: A South African Story, 1860–1914* (Durban: Madiba Publishers, 2007); Marina Carter, *Voices from Indenture: Experiences of Indian Migrants in the British Empire* (Leicester: Leicester University Press, 1996); on prisoners, see Clare Anderson, *Convicts in the Indian Ocean: Transportation from South Asia to Mauritius, 1815–1853* (London: Macmillan, 2000); Clare Anderson, *Legible Bodies: Race, Criminality and Colonialism in South Asia* (Oxford: Berg, 2004); on sailors, see Ravi Ahuja, "Networks of Subordination—Networks of the Subordinated: The Ordered Spaces of South Asian Maritime Labour in an Age of Imperialism, c. 1890-1947," in *The Limits of British Colonial Control in South Asia: Spaces of Disorder in the Indian Ocean Region,* ed. Ashwini Tambe and Harald Fischer-Tiné (London: Routledge, 2009), 13–48; G. Balachandran, "Circulation through Seafaring: Indian Seamen, 1890-1945," in *Society and Circulation: Mobile People and Itinerant Cultures in South Asia 1750–1950,* ed. Claude Markovits, Jacques Pouchepadass, and Sanjay Subrahmanyam (Delhi: Permanent Black, 2003), 89–130; Janet J. Ewald, "Crossers of the Sea: Slaves, Freedmen, and Other Migrants in the Northwestern Indian Ocean, c. 1750-1914," *American Historical Review* 105 (2000): 69–92; on pilgrims, see Radhika Singha, "Passport, Ticket, and India-rubber Stamp: 'The Problem of the Pauper Pilgrim' in Colonial India, c. 1882-1925," in *Limits of British Colonial Control,* 49–83; on merchants and trading diasporas, see Patricia Risso, *Merchants and Faith: Muslim Commerce and Culture in the Indian Ocean* (Boulder, Colo.: Westview Press, 1995). For a general overview, see

more specifically, Engseng Ho, *The Graves of Tarim: Genealogy and Mobility across the Indian Ocean* (Berkeley: University of California Press, 2006) (Hadramis); Claude Markovits, *The Global World of Indian Merchants, 1750–1947: Traders of Sind from Bukhara to Panama* (Cambridge: Cambridge University Press, 2000) (Sindhis); Goolam Vahed, "Passengers, Partnerships, and Promissory Notes: Gujarati Traders in Colonial Natal, 1870–1920," *International Journal of African Historical Studies* 38, no. 3 (2005): 449–479 (Gujaratis); on Muslim evangelizers, see Nile Green, *Bombay Islam: The Religious Economy of the West Indian Ocean, 1840–1915* (New York: Cambridge University Press, 2011).

10. J. Soske, "'Wash Me Black Again': African Nationalism, the Indian Diaspora, and Kwa-Zulu Natal, 1944–1960" (PhD dissertation, University of Toronto, 2009), 194; Green, *Bombay Islam;* Kai Kresse and Edward Simpson, eds., *Struggling with History: Islam and Cosmopolitanism in the Western Indian Ocean* (New York: Columbia University Press, 2008).

11. Green, *Bombay Islam.*

12. Sugata Bose, *A Hundred Horizons: The Indian Ocean in an Age of Global Imperialism* (Cambridge, Mass.: Harvard University Press, 2005); Mark Ravinder Frost, "'Wider Opportunities': Religious Revival, Nationalist Awakening and the Global Dimension in Colombo, 1870–1920," *Modern Asian Studies* 36, no. 4 (2002): 937–967; T. N. Harper, "Empire, Diaspora and the Languages of Globalism, 1850–1914," in *Globalization in World History,* ed. A. G. Hopkins (London: Pimlico, 2002), 141–166; Ho, *Graves of Tarim.* In addition, for overviews of Indian Ocean history, see K. N. Chaudhuri, *Trade and Civilisation in the Indian Ocean: An Economic History from the Rise of Islam to 1750* (Cambridge: Cambridge University Press, 1985); Ashin Das Gupta, *India and the Indian Ocean World: Trade and Politics,* two books: *Malabar in Asian Trade 1740–1800* (1967) and *Indian Merchants and the Decline of Surat, c. 1700–1750* (1979) (New Delhi: Oxford University Press, 2004); Michael Pearson, *The Indian*

Ocean (London: Routledge, 2003); Auguste Toussaint, *History of the Indian Ocean,* trans. June Guicharnaud (London: Routledge and Kegan Paul, 1966); M. P. M. Vink, "Indian Ocean Studies and the 'New Thalassology,'" *Journal of Global History* 2 (2007): 41–62. On port cities, see Michael N. Pearson, *Port Cities and Intruders: The Swahili Coast, India, and Portugal in the Early Modern Era* (Baltimore, Md.: Johns Hopkins University Press, 1998); Kenneth McPherson, "Port Cities as Nodal Points of Change: The Indian Ocean, 1890–1920s," in *Modernity and Culture: From the Mediterranean to the Indian Ocean,* ed. Leila Tarazi Fawaz and C. A. Bayly (New York: Columbia University Press, 2002), 75–94. For my own overview of the historiography, see Isabel Hofmeyr, Preben Kaarsholm, and Bodil Folke Frederiksen, "Introduction: Print Cultures, Nationalisms and Publics of the Indian Ocean," *Africa* 81, no. 1 (2011): 1–22; Isabel Hofmeyr, "The Black Atlantic Meets the Indian Ocean: Forging New Paradigms of Transnationalism in the Global South—Literary and Cultural Perspectives," *Social Dynamics* 33, no. 2 (2007): 3–32; Isabel Hofmeyr, "Universalizing the Indian Ocean," *PMLA* 125, no. 3 (2010): 721–729.

13. For the general idea of Indian Ocean universalism, see Bose, *Hundred Horizons;* Frost, "'Wider Opportunities,'" who also deals with pan-Buddhist movements and theosophy in "In Search of Cosmopolitan Discourse: A Historical Journey across the Indian Ocean from Singapore to South Africa, 1870–1920," in *Eyes across the Water: Navigating the Indian Ocean,* ed. Pamila Gupta, Isabel Hofmeyr, and Michael Pearson (Pretoria: University of South Africa Press, 2010), 75–108. On the other movements, see Tony Ballantyne, *Between Colonialism and Diaspora: Sikh Cultural Formations in an Imperial World* (Delhi: Permanent Black, 2006) (on Sikh transnationalism); Sukanya Banerjee, *Becoming Imperial Citizens: Indians in the Late-Victorian Empire* (Durham, N.C.: Duke University Press, 2010) (on imperial citizenship); Anne K. Bang, *Sufis and Scholars of the Sea: Family Networks in East Africa, 1860–1925* (London: Routledge-Curzon, 2003); on white laborism, see Jonathan Hyslop,

"The Imperial Working Class Makes Itself 'White': White Labourism in Britain, Australia, and South Africa before the First World War," in *The New Imperial Histories Reader,* ed. Stephen Howe (London: Routledge, 2010), 255–270; on African nationalism as a transnational phenomenon, see Christopher J. Lee, "Tricontinentalism in Question: The Cold War Politics of Alex La Guma and the African National Congress," in *Making a World after Empire: The Bandung Moment and Its Political Afterlives,* ed. Christopher J. Lee (Athens: Ohio University Press, 2010), 266–286.

14. Bose, *Hundred Horizons;* Frost, "'Wider Opportunities'"; on the Mascarenes, see Pier M. Larson, *Ocean of Letters: Language and Creolization in an Indian Ocean Diaspora* (Cambridge: Cambridge University Press, 2009).

15. Bayly, "Imagining 'Greater India.'"

16. Rochelle Pinto, *Between Empires: Print and Politics in Goa* (New Delhi: Oxford University Press, 2007).

17. There is a substantial body of work on Gandhi in South Africa, but until recently political and academic obstacles have skewed the field somewhat. The most prominent impediment has been apartheid itself, which prevented scholars from outside using archival sources in South Africa. On the academic front, national and area studies historiographies tended to misalign the story of Gandhi in South Africa and India in teleological and/or nationalist ways. Gandhi's South African experience hence becomes a simple prelude to a longer epic of Indian nationalism. Another fallout of such nationalist approaches has been the tendency to read Gandhi's ideas on "Indian-ness" as necessarily national, and to subsume any support for his vision into the same framework. As many have pointed out, such assumptions obscure the inspiration of the indentured workers strike of 1913, which was not nationalist at all. See Radhika Mongia, "Gender and the Historiography of Satyagraha in South Africa," *Gender and History* 18, no. 1 (2006): 130–149; Surendra Bhana and Goolam Vahed, *The Making of a Political Reformer: Gandhi in South Africa, 1893–1914* (New

Delhi: Manohar, 2005), 112–132; Desai and Vahed, *Inside Indenture,*
337–398. With the transnational turn in the academy and the
demise of the apartheid state, the possibilities for a more nuanced
view of Gandhi's South African context have started to emerge.
This work understands Gandhi's southern African years as a
period of intense transnational experiment that took shape
between a variety of forms of sovereignty that persisted within
empire, and on its fringes. These experiments produced different
forms of Indian-ness, many of which have subsequently fallen
from view as the nation-state and nationalist historiographies
became universalized. Gandhi's ideas of India need to be
understood less as an automatic expression of some prior
Indian-ness and more as the product of his experimental South
African years. See Bhana and Vahed, *Making of a Political Reformer;*
Keith Breckenridge, "Gandhi's Progressive Disillusionment:
Thumbs, Fingers, and the Rejection of Scientific Modernism in
Hind Swaraj," *Public Culture* 23, no. 2 (2011): 331–348; Jonathan
Hyslop, "Gandhi, Mandela and the African Modern," in *Johannes-
burg: The Elusive Metropolis,* ed. Achille Mbembe and Sarah Nuttall
(Durham, N.C.: Duke University Press, 2008), 119–136; Robert A.
Huttenback, *Gandhi in South Africa: British Imperialism and the
Indian Question, 1860–1914* (Ithaca, N.Y.: Cornell University Press,
1971); Eric Itzkin, "The Transformation of Gandhi Square: The
Search for Socially Inclusive Heritage and Public Space in the
Johannesburg City Centre" (MA dissertation, University of the
Witwatersrand, 2008), 54–76; Joseph Lelyveld, *Great Soul: Mahatma
Gandhi and His Struggle with India* (New York: Knopf, 2011); Claude
Markovits, *The Un-Gandhian Gandhi: The Life and Afterlife of the
Mahatma* (New Delhi: Permanent Black, 2003); Mongia, "Gender";
Nalini Natarajan, "Atlantic Gandhi, Caribbean Gandhi," *Economic
and Political Weekly* 18 (2009): 43–52; Anthony Parel, "The Origins
of *Hind Swaraj,*" in *Gandhi and South Africa: Principles and Politics,* ed.
Judith M. Brown and Martin Prozesky (Pietermaritzburg:

University of Natal Press, 1996), 35–66; Maureen Swan, *Gandhi: The South African Experience* (Johannesburg: Ravan Press, 1985). There is a further distinction between the scholarship emerging from South Africa and that from India; the former tends to treat Gandhi as secular and transnational, the latter as nonsecular and national. This work attempts to crosshatch these traditions of scholarship.

18. Anil Nauriya, *The African Element in Gandhi,* http://www.sahistory.org.za/sites/default/files/TheAfricanElementinGandhi%20by%20Anil%20Nauriyafinal.pdf (accessed June 18, 2012). For Gandhi's views on Africans, see J. H. Stone, "M. K. Gandhi: Some Experiments with Truth," *Journal of Southern African Studies* 16, no. 4 (1990): 721–740.

19. Soske, "'Wash Me Black Again,'" 89–101, 159–163.

20. Heather Hughes, *The First President: A Life of John L. Dube, Founding President of the ANC* (Johannesburg: Jacana, 2011), 111.

21. Tony Ballantyne, *Orientalism and Race: Aryanism and the British Empire* (Basingstoke: Palgrave, 2002), 17.

22. Patrick Eisenlohr, *Little India: Diaspora, Time and Ethnolinguistic Belonging in Hindu Mauritius* (Berkeley: University of California Press, 2006), 229–232; Ravindra K. Jain, "Indian Diaspora, Globalisation and Multiculturalism: A Cultural Analysis," in *Tradition, Pluralism and Identity: In Honour of TN Madan,* ed. Veena Das, Dipankar Gupta, and Patricia Uberoi (New Delhi: Sage, 1999), 195–218; Makarand Paranjape, ed., *In Diaspora: Theories, Histories, Texts* (New Delhi: Indialog Publications, 2001), 1–5.

23. Frederick Cooper and Ann Laura Stoler, eds., *Tensions of Empire: Colonial Cultures in a Bourgeois World* (Berkeley: University of California Press, 1997).

24. David M. Henkin, *City Reading: Written Words and Public Spheres in Antebellum New York* (New York: Columbia University Press, 1998), 101–106.

25. Anderson, *Imagined Communities,* 34, 37.

26. Christopher A. Reed, *Gutenberg in Shanghai: Chinese Print Capitalism 1876–1937* (Honolulu: University of Hawai'i Press, 2004), 9.

27. Sheldon Pollock, *The Language of the Gods in the World of Men: Sanskrit, Culture, and Power in Premodern India* (Berkeley: University of California Press, 2006), 558.

28. Karin Barber, "Audiences and the Book," *Current Writing* 13, no. 2 (2001): 9–19.

29. Ellen Gruber Garvey, "Scissorizing and Scrapbooks: Nineteenth-Century Reading, Remaking, and Recirculating," in *New Media 1740–1915,* ed. Lisa Gitelman and Geoffrey B. Pingree (Cambridge, Mass.: MIT Press, 2003), 207–227; Ross Harvey, "Bringing the News to New Zealand: The Supply and Control of Overseas News in the Nineteenth Century," *Media History* 8, no. 1 (2002): 21–34; Richard B. Kielbowicz, "Newsgathering by Printers' Exchanges before the Telegraph," *Journalism History* 9, no. 2 (1982): 42–48.

30. Ramananda Chatterjee, "Origin and Growth of Journalism among Indians," *Annals of the American Academy of Political and Social Science* 145 (1929): 161–168; Milton Israel, *Communications and Power: Propaganda and the Press in the Indian Nationalist Struggle, 1920–1947* (Cambridge: Cambridge University Press, 1994), 99–151; Simon J. Potter, *News and the British World: The Emergence of an Imperial Press System* (Oxford: Oxford University Press, 2003), 16.

31. Pandiri, *A Comprehensive, Annotated Bibliography,* 168, 183, 204, 213, 229.

32. Deep Kanta Lahiri Choudhury, *Telegraphic Imperialism: Crisis and Panic in the Indian Empire, c. 1830–1920* (London: Palgrave Macmillan, 2010); Saumendranath Bera, "Confronting the Colonial Order: Gandhiji, the *Green Pamphlet* and Reuter," in *Webs of History: Information, Communication, and Technology from Early to Post-colonial India,* eds. Amiya Kumar Bagchi, Dipankar Sinha, and Barnita Bagchi (New Delhi: Manohar, 2005), 209–230.

33. Pyarelal, *The Discovery of Satyagraha: On the Threshold,* vol. 2 (Bombay: Sevak Prakasha, 1980), 3–15.

34. Kelly J. Mays, "The Disease of Reading and Victorian Periodicals," in *Literature in the Marketplace: Nineteenth-Century British Publishing and Reading Practices,* ed. John O. Jordan and Robert L. Patten (Cambridge: Cambridge University Press, 1995), 165-194.

35. Leah Price, *How to Do Things with Books in Victorian Britain* (Princeton: Princeton University Press, 2012).

36. Faith Binckes, "Lines of Engagement: *Rhythm,* Reproduction, and the Textual Dialogues of Early Modernism," in *Little Magazines and Modernism: New Approaches,* ed. Suzanne W. Churchill and Adam McKible (Aldershot: Ashgate, 2007), 21-34; Mays, "Disease of Reading"; Leah Price, *The Anthology and the Rise of the Novel: From Richardson to George Eliot* (Cambridge: Cambridge University Press, 2000), e-book.

37. Linda K. Hughes and Michael Lund, *The Victorian Serial* (Charlottesville: University Press of Virginia, 1991); Suzanne W. Churchill and Adam McKible, eds., "Introduction," *Little Magazines and Modernism: New Approaches* (Aldershot: Ashgate, 2007), 1-18; Binckes, "Lines of Engagement"; Price, *Anthology.*

38. Hughes and Lund, *Victorian Serial;* for an excellent exception, see Parel, "The Origins."

39. Desai and Vahed, *Inside Indenture,* 376-398; Mongia, "Gender"; Swan, *Gandhi,* 245-256.

40. Alon Confito and Ajay Skaria, "The Local Life of Nationhood," *National Identities* 4, no. 1 (2002): 7-24; Uday Singh Mehta, "Gandhi and the Common Logic of War and Peace," *Raritan* 30, no. 1 (2010): 134-156; Mehta, "Patience"; Ajay Skaria, "Gandhi's Politics: Liberalism and the Question of the Ashram," *South Atlantic Quarterly* 101, no. 4 (2002): 954-986; Tridip Suhrud, "Hind Swaraj," in *Ten Books that Changed the British Empire,* ed. Antoinette Burton and Isabel Hofmeyr (Durham, N.C.: Duke University Press, forthcoming).

41. Mehta, "Gandhi", 25.

42. Confito and Skaria, "Local Life," 19.

43. Quoted in Confito and Skaria, "Local Life," 20.

44. This section is drawn from the material discussed in note 17.

1. Printing Cultures in the Indian Ocean World

1. This information surfaced in an ongoing legal wrangle between two descendants of the ship's original owners, the brothers Abdul and Abdullah Zaveri of Dada Abdullah and Co. See "Southern News—Tamil Nadu," www.newindpress.com (accessed September 28, 2007); "Ship Ownership Row," *Deccan Herald*, March 14, 2004, http://www.deccanherald.com/archives/mar142004/s17.asp (accessed January 12, 2006); "Choppy Waters Ahead for Gandhi's Ship," *The Times of India*, October 5, 2006, http://timesofindia.indiatimes.com/articleshow/2099998.coms (accessed January 12, 2007).

2. Pyarelal, *The Birth of Satyagraha: From Petitioning to Passive Resistance*, vol. 3 (Ahmedadabad: Navajivan Publishing House, 1986), 63–65.

3. M. K. Gandhi, *An Autobiography or A Story of My Experiments with Truth*, trans. Mahadev Desai (Bombay: Navajivan, n.d.), 302–304, chap. 20, e-book; M. K. Gandhi, *Satyagraha in South Africa* (Ahmedabad: Navajivan, 1928).

4. On peaceful cosmopolitan trade, see Amitav Ghosh, *In an Antique Land: History in the Guise of a Traveler's Tale* (London: Vintage, 1992); on how Grotius's idea of *mare liberum* both draws on and erases Indian Ocean politics, see Stephanie Jones, "The Poetic Ocean in Mare Liberum," in *Law and Art: Justice, Ethics and Aesthetics*, ed. Oren Ben-Dor (Abdingdon: Routledge, 2010), 188–203; on island utopias (and dystopias) in the Indian Ocean, see Stephanie Jones, "Colonial to Postcolonial Ethics: Indian Ocean 'Belongers,' 1668-2008," *Interventions* 11, no. 2 (2009): 212–234, and Pamila Gupta, "Island-ness in the Indian Ocean," in *Eyes across the Water: Navigating the Indian Ocean*, ed. Pamila Gupta, Isabel Hofmeyr, and Michael Pearson (Pretoria: University of South Africa Press, 2010), 275–285; on pirate republics, see Jan Rogoziński, *Honor among*

Thieves: Captain Kidd, Henry Every, and the Pirate Democracy in the Indian Ocean (Mechanicsburg: Stackpole Books, 2000); on Afro-Asian solidarity, see Christopher J. Lee, ed., *Making a World after Empire: The Bandung Moment and Its Political Afterlives* (Athens: Ohio University Press, 2010), and Vijay Prasad, *The Darker Nations: A People's History of the Third World* (New York: New Press, 2007); also see Seng Tan and Amitav Acharya, eds., *Bandung Revisited: The Legacy of the 1955 Asian African Conference for International Order* (Singapore: National University of Singapore Press, 2008); on theosophy, see Mark Ravinder Frost, "In Search of Cosmopolitan Discourse: A Historical Journey across the Indian Ocean from Singapore to South Africa, 1870-1920," in *Eyes across the Water,* 75-108; on Greater India, see Susan Bayly, "Imagining 'Greater India': French and Indian Visions of Colonialism in the Indic Mode," *Modern Asian Studies* 38, no. 3 (2004): 703-744.

5. Ghosh, *Antique Land.*

6. Amitav Ghosh, "Confessions of a Xenophile," *Chimurenga* 14 (2009): Cluster A, 35-41.

7. Transnational book history is now an extensive field: relevant here are Bill Bell, "Crusoe's Books: The Scottish Emigrant Reader in the Nineteenth Century," in *Across Boundaries: The Book in Culture and Commerce,* ed. Bill Bell, Philip Bennett, and Jonquil Bevan (Delaware: Oak Knoll Press, 2000), 116-129; Rimi B. Chatterjee, *Empires of the Mind: A History of the Oxford University Press in India under the Raj* (New Delhi: Oxford University Press, 2006); Robert Fraser, *Book History through Postcolonial Eyes: Rewriting the Script* (London: Routledge, 2008); Priya Joshi, *In Another Country: Colonialism, Culture, and the English Novel in India* (New Delhi: Oxford University Press, 2002); Martyn Lyons and John Arnold, eds., *A History of the Book in Australia 1891–1945: A National Culture in a Colonised Market* (St. Lucia: University of Queensland Press, 2001); James Raven, *London Booksellers and American Customers: Transatlantic Literary Community and the Charleston Library Society, 1748–1811* (Columbia: University of South Carolina Press, 2002). See Juan

R. I. Cole, "Printing and Urban Islam in the Mediterranean World, 1890-1920," in *Modernity and Culture: From the Mediterranean to the Indian Ocean,* ed. Leila Tarazi Fawaz and C. A. Bayly (New York: Columbia University Press, 2002); Nile Green, *Bombay Islam: The Religious Economy of the West Indian Ocean, 1840–1915* (New York: Cambridge University Press, 2011); Nile Green, "Journeymen, Middlemen: Travel, Transculture, and Technology in the Origins of Muslim Printing," *International Journal of Middle Eastern Studies* 41 (2009): 203-224; Nile Green, "Persian Print and the Stanhope Revolution: Industrialization, Evangelicalism, and the Birth of Printing in Early Qajar Iran," *Comparative Studies of South Asia, Africa and the Middle East* 30, no. 3 (2010): 473-490; Nile Green, "Saints, Rebels and Booksellers: Sufis in the Cosmopolitan Western Indian Ocean, c. 1780-1920," in *Struggling with History: Islam and Cosmopolitanism in the Western Indian Ocean,* ed. Edward Simpson and Kai Kresse (New York: Columbia University Press, 2008); and Francis Robinson, "Technology and Religious Change: Islam and the Impact of Print," *Modern Asian Studies* 27, no. 1 (1993): 229-251, on transnational Muslim printing traditions. Although national in focus, book history in India is also relevant to this study. See Abhijit Gupta and Swapan Chakravorty, eds., *Print Areas: Book History in India* (New Delhi: Permanent Black, 2004); Abhijit Gupta and Swapan Chakravorty, eds., *Moveable Type: Book History in India* (New Delhi: Permanent Black, 2008); Francesca Orsini, *The Hindi Public Sphere (1920–1940): Language and Literature in the Age of Nationalism* (New York: Oxford University Press, 2002). On newspapers and periodicals in empire, see Milton Israel, *Communications and Power: Propaganda and the Press in the Indian Nationalist Struggle, 1920–1947* (Cambridge: Cambridge University Press, 1994); Simon J. Potter, *News and the British World: The Emergence of an Imperial Press System* (Oxford: Oxford University Press, 2003); Chandrika Kaul, *Reporting the Raj: The British Press and India, 1880–1922* (Manchester: Manchester University Press, 2003); Chandrika Kaul, ed., *Media and the British Empire* (Basingstoke: Palgrave Macmillan, 2006); David Finkelstein and Douglas M. Peers, eds., *Negotiating India in the Nineteenth-Century Media* (London: Palgrave Macmillan, 2000); Julie F. Codell, "Introduction: The Nineteenth-Century News from India," *Victorian Periodicals Review* 37, no. 2 (2004): 106–123; Julie F. Codell, "Getting the Twain to Meet: Global Regionalism in 'East and West': A Monthly Review," *Victorian Periodicals Review* 37, no. 2 (2004): 214–232; Sukeshi Kamra, *The Indian Periodical Press and the Production of Nationalist Rhetoric* (London: Palgrave Macmillan, 2011); J. Don Vann and Rosemary T. van Arsdel, eds., *Periodicals of Queen Victoria's Empire: An Exploration* (Toronto: University of Toronto Press, 1966).

8. Raven, *London Booksellers;* Kaul, *Media;* Sydney Shep, "Mapping the Migration of Paper: Historical Geography and New Zealand Print Culture," in *The Moving Market,* ed. Peter Isaac and Barry Mackay (Delaware: Oak Knoll Press, 2001), 179–192; see also Potter, *News;* Finkelstein and Peers, *Negotiating India;* Kaul, *Reporting the Raj.* One exception to this trend is Israel (*Communications and Power,* 246–316), who covers India and "The Struggle Overseas."

9. Isabel Hofmeyr, Preben Kaarsholm, and Bodil Folke Frederiksen, "Introduction: Print Cultures, Nationalisms and Publics of the Indian Ocean," *Africa* 81, no. 1 (2011): 1–22.

10. On Muslim printing, see Green, *Bombay Islam;* on Africans in the printing sector in South Africa, see Jeffrey Peires, "Lovedale Press: Literature for the Bantu Revisited," *English in Africa* 7, no. 1 (1980): 71–85; Tim White, "The Lovedale Press during the Directorship of R. H. W. Shepherd, 1930–1955," *English in Africa* 19, no. 2 (1992): 69–84; on printing in India, see below, notes 13–15; on Mauritius, see Auguste Toussaint, "Mauritius, Réunion, Madagascar and the Seychelles," in *The Spread of Printing: Eastern Hemisphere,* ed. Colin Clair (Amsterdam: Vangendt; London: Routledge & Kegan Paul; New York: Abner Schram, 1969), 49–52; on East Africa, see James R. Brennan, "Politics and Business in the Indian Newspapers of Colonial Tanganyika," *Africa* 81, no. 1 (2011): 42–67; Bodil Folke

Frederiksen, "'The Present Battle Is the Brain Battle': Writing and Publishing a Kikuyu Newspaper in the Pre-Mau Mau Period in Kenya," in *Africa's Hidden Histories: Everyday Literacy and the Making of the Self*, ed. Karin Barber (Bloomington: Indiana University Press, 2006), 278–313; Bodil Folke Frederiksen, "Print, Newspapers and Audiences in Colonial Kenya: African and Indian Improvement, Protest and Connections," *Africa* 81, no. 1 (2011): 155–172; Wangari Muoria-Sal, Bodil Folke Frederiksen, John Lonsdale, and Derek Peterson, *Writing for Kenya: The Life and Works of Henry Muoria* (Leiden: Brill, 2009); on South Africa, see Thembisa Waetjen and Goolam Vahed, "The Diaspora at Home: Indian Views and the Making of Zuleikha Mayat's Public Voice," *Africa* 81, no. 1 (2011): 23–41.

11. Green, *Bombay Islam*.

12. For research on British printers, see SAPPHIRE (Scottish Archive of Print and Publishing History Records), www.sapphire.ac.uk/index.htm (accessed September 1, 2012); for the construction of whiteness in the dominions, see Jonathan Hyslop, "How the British Working Class Became White: The Symbolic (Re)formation of Racialized Capitalism," *Journal of Historical Sociology* 11, no. 3 (1999): 316–340.

13. *Indian Printers' Journal*, January 1912, 2; Green, *Bombay Islam*, 95; Francesca Orsini, *Print and Pleasure: Popular Literature and Entertaining Fictions in Colonial North India* (Delhi: Permanent Black, 2009), 38; Ulrike Stark, *An Empire of Books: the Naval Kishore Press and the Diffusion of the Printed Word in Colonial India* (New Delhi: Oxford University Press, 2007).

14. On state printing, see Lal Chand & Sons, *State-Owned Printing Presses and Their Competition with Private Trade* (Calcutta: Lal Chand & Sons, 1923); Anant Kakba Priolkar, *The Printing Press in India: Its Beginnings and Early Development Being a Qua[r]tercentenary Commemoration Study of the Advent of Printing in India (in 1556)* (Bombay: Marathi Shamshodhana Mandala, 1956), 87; on Christian mission presses, see Anindita Ghosh, "Between Text and Reader: The Experience of Christian Missionaries in Bengal, 1800–1850," in

Free Print and Non-Commercial Publishing since 1700, ed. James Raven (London: Ashgate, 2000), 162–176; J. Mangamma, *Book Printing in India: With Special Reference to the Contribution of European Scholars to Telegu (1746–1857)* (Nellore: Bangorey Books, 1975), 51; Ellen E. McDonald, "The Modernizing of Communication: Vernacular Publishing in Nineteenth Century Maharashtra," *Asian Survey* 8, no. 7 (1968): 589–606; Priolkar, *Printing Press,* 80; Robinson, "Technology"; Graham W. Shaw, "Communications between Cultures: Difficulties in the Design and Distribution of Christian Literature in Nineteenth-Century India," in *The Church and the Book,* ed. R. N. Swanson (Woodbridge: Boydell Press, 2004), 339–356; Graham W. Shaw, "The Cuttack Mission Press and Early Oriya Printing," *British Library Journal* 3, no. 1 (1977): 29–43; Graham W. Shaw, "Printing at Mangalore and Tellicherry by the Basel Mission," *Libri* 27, no. 2 (1977): 154–164; on the spread of the iron press, see Green, "Persian Print"; on lithography, see Stark, *Empire of Books,* 45–49; Graham W. Shaw, "Calcutta: Birthplace of the Indian Lithographed Book," *Journal of the Printing Historical Society* 27 (1998): 89–111; on the paper industry, see *Imperial Gazetteer of India: The Indian Empire: Economic,* vol. 3 (Oxford: Clarendon Press, 1908), 206; Stark, *Empire of Books,* 189–190.

15. Ulrike Stark, "Hindi Publishing in the Heart of an Indo-Persian Cultural Metropolis: Lucknow's Newal Kishore Press (1858–1895)," in *India's Literary History: Essays on the Nineteenth Century,* ed. Stuart Blackburn and Vasudha Dalmia (New Delhi: Permanent Black, 2004), 251–279; Stark, *Empire of Books.*

16. Mark Ravinder Frost, "Asia's Maritime Networks and the Colonial Public Sphere, 1840–1920," *New Zealand Journal of Asian Studies* 6, no. 2 (2004): 5–36.

17. Somerset Playne, comp., *Southern India: Its History, People, Commerce, and Industrial Resources* (London: The Foreign and Colonial Compiling and Publishing Company, 1914–1915), 638–642, 654–657.

18. Ghosh, "Between Text and Reader"; Anindita Ghosh, "Cheap Books, 'Bad' Books: Contesting Print-Cultures in Colonial

Bengal," *South Asia Research* 18, no. 2 (1998): 173–194; Anindita Ghosh, *Popular Publishing and the Politics of Language and Culture in a Colonial Society 1778–1905* (New Delhi: Oxford University Press, 2006); also see Tapti Roy, "Disciplining the Printed Text: Colonial and Nationalist Surveillance of Bengali Literature," in *Texts of Power: Emerging Disciplines in Colonial Bengal*, ed. Partha Chatterjee (Minneapolis and London: University of Minnesota Press, 1995), 30–62.

19. T. Fisher, *The Elements of Letterpress Printing, Composing and Proof- reading: A Practical Manual for Indian Artisans* (Madras: Higginbotham & Co., 1906), vi; *Imperial Gazetteer of India: The Indian Empire: Administrative*, vol. 4 (Oxford: Clarendon Press, 1909), 321.

20. Veena Naregal, *Language Politics, Elites and the Public Sphere* (New Delhi: Permanent Black, 2001), 179–180; Stark, *Empire of Books*, 79.

21. Bernard Ellis, "Christian Literature, Baptist Mission Press and the Serampore Legacy," http://www.wmward.org/Bernard %20Ellis%20html/bellis1.html (accessed January 23, 2012).

22. See Stark's account of Naval Kishore (*Empire of Books*, 107–163); also see biographical accounts of G. A. Natesan in *Indian Review* 27, no. 1 (1925). Natesan was editor and founder of *Indian Review*, an important model for *Indian Opinion*. Gandhi met Natesan in 1896, and they remained in close contact. Another reforming editor with a powerful interest in the diaspora was Ramananda Chatterjee, who established the "diaspora-friendly" *Modern Review*. For biographical accounts, see R. K. Dasgupta, "Ramananda Chatterjee," in *Some Eminent Indian Editors* (New Delhi: Ministry of Information and Broadcasting, n.d.), 123–141. Printing presses themselves were also named to reflect ideals of social progress. See the Reform Press, *Indian Social Reformer*, July 5, 1896. See also *Indian Opinion*'s profiling of editors who formed an important constituency in the South Africa-India support networks, in "South Africa's Voice in India," October 15, 1910. Benarsidas Chaturvedi was another

important reforming editor who devoted much of his life to the cause of "Indians overseas."

23. For Indian printing in South Africa, see C. G. Henning, *The Indentured Indian in Natal (1860–1917)* (New Delhi: Promilla & Co., 1993), 184–189; Uma Dhupelia-Mesthrie, *Gandhi's Prisoner? The Life of Gandhi's Son Manilal* (Cape Town: Kwela, 2004); Waetjen and Vahed, "Diaspora at Home." On East Africa, see Frederiksen, "Present Battle" and "Print"; Muoria-Sal et al., *Writing for Kenya;* Fay Gadsden, "The African Press in Kenya, 1945-1952," *Journal of African History* 21, no. 4 (1980): 515-535; James Ogude, "The Vernacular Press and the Articulation of Luo Ethnic Citizenship: The Case of Achieng' Oneko's *Ramogi,*" *Current Writing* 13, no. 2 (2001): 42–55; James F. Scotton, "Kenya's Maligned African Press: A Reassessment" (paper presented at the Annual Meeting of the Association of Education for Journalism, San Diego, California, August 18-21, 1974), www.eric.ed.gov/PDFS/ED096679.pdf (accessed September 1, 2012).

24. Toussaint, "Mauritius."

25. NAD IRD 18 543/1903. "P. S. Aiyar, Colonial Indian News, Durban: Asks for a Permit to Bring an Indian Compositor from Mauritius to Durban."

26. Green, *Bombay Islam,* 24, 97–99, 118-154, 208-234.

27. The Telegu Baptist Home Missionary Society, which arose out of the American Baptist Foreign Mission Society, sent out missionaries to Natal to work among indentured laborers. See John Rungiah, *The First and Second Annual Reports of the Telegu Baptist Mission, Natal, South Africa, 1903–1905* (Madras: ME Press, 1905); Robert G. Torbet, *Venture of Faith: The Story of the ABFMS and the Woman's ABFMS* (Philadelphia: Judson Press, 1955), 253-270.

28. Pyarelal, *The Discovery of Satyagraha: On the Threshold,* vol. 2 (Bombay: Sevak Prakasha, 1980), 4–5; A. G. Downes, *Printer's Saga: Being a History of the South African Typographical Union* (Johannesburg: South African Typographical Union, 1952), 13.

29. L. E. Neame, *The Asiatic Danger in the Colonies* (London: George Routledge & Sons, 1907), 71.

30. See advertisements in *Indian Printers' Journal* and Playne, *Southern India*, 133-135, 184-185.

31. Pyarelal, *Birth*, 66, for journalists wanting to assist Gandhi; for the opposite, see material on Neame in Chapter 4.

32. Fisher, *Elements*, 427.

33. *Natal Mercury*, November 30, 1898; NAD PM 28 1902/1074. "Secretary, South African Typographical Union, Durban. Forwards Copy of Petition Adopted at the Recent Conference Held in Durban re Imposition of 100% Duty upon all Commercial Printed Matter and Stationery from Over Sea Ports"; Toussaint, "Mauritius," 49.

34. Edward Conner, "Printing in the East: India: The Native Printer—At Play," *South African Printer and Stationer*, January 1927: 34-35; Fisher, *Elements*, 126, 378.

35. Fisher, *Elements*, 185, 261.

36. Ibid., 86, 101, 59.

37. Peter Francis, *Print and Politics: A History of Trade Unions in the New Zealand Printing Industry 1865-1995* (Wellington: Victoria University Press, 2001), 18.

38. Peires, "Lovedale Press"; White, "Lovedale Press."

39. B. Willan, *Sol Plaatje: South African Nationalist, 1876-1932* (Johannesburg: Ravan Press, 1985), 158-159.

40. Frederiksen, "Present Battle" and "Print"; Muoria-Sal et al., *Writing for Kenya;* Ogude, "Vernacular Press."

41. Sana Aiyar, "Print, Newspapers and Audiences in Colonial Kenya: African and Indian Improvement, Protest and Connection," *Africa* 81, no. 1 (2011): 132-154.

42. Waetjen and Vahed, "Diaspora at Home," 25.

43. Dhupelia-Mesthrie, *Gandhi's Prisoner*, 156.

44. For Beira, see advertisements in *Beira News and East Coast Chronicle*, September 4, 1917, and *Beira Post*, March 23, 1898 (advertisements on Beira Printing and Publishing Works); on Dar-es-Salaam, see

Brennan, "Politics and Business"; on Zanzibar, see Martin Sturmer, *The Media History of Tanzania,* www.tanzania.at/mht/The _Media_History_of_Tanzania.pdf (accessed November 1, 2012), 275–277; on Mombasa, see Zarina Patel, *Challenge to Colonialism: The Struggle of Alibhai Mulla Jeevanjee for Equal Rights in Kenya* (Nairobi: n.p., 1997), 30–36; on Nairobi, see Frederiksen, "Present Battle" and "Print"; Muoria-Sal et al., *Writing for Kenya.*

45. Letterheads: NAD CSO 1848 1907/8564. "International Printing Press, and Others. Applications for, or Renewal of Licenses for Newspapers, Notaries, Dentists, Chemists and Medical Practitioners, 1908"; applications: NAD CSO 1848 1907/8564; SAD JUS 360 4/188/22 12/6/22. "Registration of 'Hindi' as a Newspaper. Natal Province"; for colonial state surveillance on *Indian Opinion* NAD II 1/128 I1296/1904. "Minute Paper. Comment of 'Indian Opinion' on the Report of the Protector of Indian Immigrants from the Year 1903"; NAD II 1/133 I164/1905. "Protector of Indian Immigrants, Durban: 'Indian Opinion,' January 21, 1905. Further Reference to Assault Case Ramsay Collieries"; SAD PM 42 1903/1767. "Article 'Indian Opinion.' "

46. Zarina Patel, *Unquiet: The Life and Times of Makhan Singh* (Nairobi: Zand Graphics, 2006), and *Challenge to Colonialism.*

47. Patel, *Unquiet,* 33–35.

48. Patel, *Challenge to Colonialism,* 30–36.

49. Hofmeyr, Kaarsholm, and Frederiksen, "Introduction," 7.

50. NAD CSO 1848 1907/8564.

51. Information on Nazar from Surendra Bhana and James D. Hunt, eds., *Gandhi's Editor: The Letters of M. H. Nazar* (New Delhi: Promilla, 1989), 3. Pyarelal (*Birth,* 69) portrays him as having more experience. On others, see Dhupelia-Mesthrie, *Gandhi's Prisoner,* 59, 70.

52. Henry S. L. Polak, "My Ten Years' Service," *Golden Jubilee Phoenix Settlement 1904–1954* (Durban: International Printing Press, 1954), 7–14; Laurel Brake, "The Old Journalism and the New: Forms of Cultural Production in London in the 1880s," in *Papers for the*

Millions: The New Journalism in Britain, 1850s to 1914, ed. Joel H.
Wiener (New York: Greenwood Press, 1988), 1–24; Joanne Shat-
tock, "Showman, Lion Hunter, or Hack: The Quarterly Editor at
Midcentury," in *Innovators and Preachers: The Role of the Editor in
Victorian England,* ed. Joel H. Wiener (New York: Greenwood
Press, 1985), 161–183; Joel H. Wiener, "Introduction," in
Innovators and Preachers: xi–xix.

53. NAD CSO 1848 1907/8564.
54. Stark, *Empire of Books,* 2.
55. Brennan, "Politics and Business."
56. *Avadh Akhbar,* the periodical of Naval Kishore Press, listed the
amounts given by different patrons, ranging from Rs 2 to Rs 23
(Stark, *Empire of Books,* 358–359; the date of the issue is 1875). Uma
Das Gupta lists circulation figures in the 1870s with some publications as low as fifty. See "The Indian Press 1870–1880: A Small
World of Journalism," *Modern Asian Studies* 11, no. 2 (1977): 213–235.
While these figures would have risen by the early 1900s, the
system of subscriber patronage would have continued to enable
periodicals with low circulations to survive.
57. Readership and population figures from Gandhi, *Satyagraha,* 92;
see also Uma Shashikant Mesthrie, "From Advocacy to Mobilization: *Indian Opinion,* 1903–1914," in *South Africa's Alternative Press:
Voices of Protest and Resistance, 1880s–1990s,* ed. Les Switzer (Cambridge:
Cambridge University Press, 1987), 104.

2. Gandhi's Printing Press

1. Pyarelal, *The Birth of Satyagraha: From Petitioning to Passive Resistance,* vol. 3 (Ahmedabad: Navajivan Publishing House, 1986), 66.
2. Pyarelal, *The Discovery of Satyagraha: On the Threshold,* vol. 2
(Bombay: Sevak Prakasha, 1980), 193–194.
3. Reynolds Stone, *The Albion Press: An Essay First Published in the
Journal of the Printing Historical Society* (London: Printing Historical
Society, 2005), 58.

4. Pyarelal, *Discovery*, 193.

5. Pyarelal, *Birth*, 434; NAD CSO 1735 1903/6053, "Magistrate, Durban, Declaration for Bond for New Newspaper, 'The Volunteer.'"

6. Pyarelal, *Birth*, 434.

7. Fatima Meer, *Portrait of Indian South Africans* (Durban: Avon House, 1969), 56–57.

8. Albert West, "In the Early Days with Gandhi," www.mkgandhi.org/articles/earlydays.htm (accessed December 1, 2011).

9. A. G. Downes, *Printer's Saga: Being a History of the South African Typographical Union* (Johannesburg: South African Typographical Union, 1952), 80.

10. Ibid., 99.

11. Surendra Bhana and James D. Hunt, eds., *Gandhi's Editor: The Letters of M. H. Nazar* (New Delhi: Promilla, 1989).

12. Ibid., 109.

13. Ibid., 112.

14. Anant Kakba Priolkar, *The Printing Press in India: Its Beginnings and Early Development Being a Qua[r]tercentenary Commemoration Study of the Advent of Printing in India (in 1556)* (Bombay: Marathi Shamshodhana Mandala, 1956), 77.

15. Bhana and Hunt, *Gandhi's Editor*, 89, 112.

16. This figure is derived from T. Fisher, *The Elements of Letterpress Printing, Composing and Proofreading: A Practical Manual for Indian Artisans* (Madras: Higginbotham & Co., 1906), 95, who reckoned on a rate of 8,000 ens per eight-hour day. Philip Gaskell, *A New Introduction to Bibliography* (London: Oxford University Press, 1972), 54, gives the same rate of 1,000 ens per hour.

17. Bhana and Hunt, *Gandhi's Editor*, 114.

18. Maureen Swan, *Gandhi: The South African Experience* (Johannesburg: Ravan Press, 1985), 57.

19. Ashwin Desai and Goolam Vahed, *Inside Indenture: A South African Story, 1860–1914* (Durban: Madiba Publishers, 2007), 307–322.

20. Maynard Swanson, "'The Asiatic Menace': Creating Segregation in Durban, 1870–1900," *International Journal of African Historical Studies* 16, no. 3 (1983): 401–421.

21. Downes, *Printer's Saga.*

22. Robert Morrell, *From Boys to Gentlemen: Settler Masculinity in Colonial Natal 1880–1920* (Pretoria: University of South Africa Press, 2001).

23. Downes, *Printer's Saga,* 16.

24. Pyarelal, *Birth,* 85–86.

25. Heather Hughes, *The First President: A Life of John L. Dube, Founding President of the ANC* (Johannesburg: Jacana, 2011), 89.

26. Heather Hughes, e-mail on the history of Dube's printing press, January 10, 2012.

27. Hughes, *First President,* 103–104.

28. Richard Gabriel Rummonds, *Nineteenth-Century Printing Practices and the Iron Handpress,* vols. 1 and 2 (Delaware and London: Oak Knoll Press and The British Library, 2004), 147.

29. Hughes, *First President,* 91.

30. Heather Hughes, "'The Coolies Will Elbow Us Out of the Country': African Reactions to Indian Immigration in the Colony of Natal, South Africa," *Labour History Review* 72, no. 2 (2007): 155–168.

31. Karen E. Flint, *Healing Traditions: African Medicine, Cultural Exchange and Competition in South Africa, 1820–1948* (Athens: Ohio University Press, 2008), 158–182; Jon Soske, "'Wash Me Black Again': African Nationalism, the Indian Diaspora, and Kwa-Zulu Natal, 1944–1960" (PhD. dissertation, University of Toronto, 2009), 59.

32. Hughes, "'Coolies,'" 159–160, 162.

33. Michael Green, "Exorcizing the Past: Voices from the Present," in *Religion and Spirituality in South Africa: New Perspectives,* ed. Duncan Brown (Pietermaritzburg: University of KwaZulu-Natal Press, 2009), 167–190.

34. Desai and Vahed, *Inside Indenture,* 423–425.

35. Nile Green, *Bombay Islam: The Religious Economy of the West Indian Ocean, 1840–1915* (New York: Cambridge University Press, 2011), 208–234; Nile Green, "Islam for the Indentured Indian: A Muslim Missionary in Colonial South Asia," *Bulletin of the School of Oriental and African Studies* 71, no. 3 (2008): 529–553.

36. J. T. F. Jordens, *Dayananda Sarasvati: His Life and Ideas* (Delhi: Oxford University Press, 1997).

37. For an international history of the Arya Samaj, see Nardev Vedalankar and Manohar Somera, *Arya Samaj and Indians Abroad* (Durban: Sarvadeshik Arya Pratinidhi Sabha, n.d.), 35, who represent the indentured diaspora as part of the history of the Samaj in the sense that the "desperate" condition of the indentured workers called forth a redemptive response. On the Samaj in South Africa, see Thillayvel Naidoo, *The Arya Samaj Movement in South Africa* (Delhi: Motilal Banarsidass, 1992).

38. Goolam Vahed, "Swami Shankeranand and the Consolidation of Hinduism in Natal, 1908–1913," *Journal for the Study of Religion* 10, no. 2 (1988): 3–35.

39. Goolam Vahed, "Constructions of Community and Identity among Indians in Colonial Natal, 1860–1910: The Role of the Muharram Festival," *Journal of African History* 43, no. 1 (2002): 77–93; Vahed, "Swami Shankeranand," 22.

40. Vahed, "Swami Shankeranand," 8.

41. Rummonds, *Nineteenth-Century Printing Practices*, 33. The measurements are recorded variously as 50 × 75 in Uma Dhupelia-Mesthrie, *Gandhi's Prisoner? The Life of Gandhi's Son Manilal* (Cape Town: Kwela, 2004), 59, and as 80 × 40 feet in West, "Early Days".

42. West, "Early Days"; also see Pyarelal, *Birth*, 437.

43. Dhupelia-Mesthrie, *Gandhi's Prisoner*, 59–60; West, "Early Days."

44. Dhupelia-Mesthrie, *Gandhi's Prisoner*, 57–60, 67–75.

45. Pyarelal, *Birth*, 435–437.

46. Dhupelia-Mesthrie, *Gandhi's Prisoner*, 69.

47. Uma Shashikant Mesthrie, "From Advocacy to Mobilization: *Indian Opinion, 1903-1914,*" in *South Africa's Alternative Press: Voices of Protest and Resistance, 1880s–1990s,* ed. Les Switzer (Cambridge: Cambridge University Press, 1987), 99-126; B. Pachai, "The History of the 'Indian Opinion' 1903-1914," in *Archives Yearbook for South African History* (Cape Town: Office of the Director of Archives, 1963), 1-127c.

48. Mesthrie, "Advocacy," 99-126.

49. M. K. Gandhi, *An Autobiography or A Story of My Experiments with Truth,* trans. Mahadev Desai (Bombay: Navajivan, n.d.), 304, 3974 of 6567, e-book.

50. Prabhudas Gandhi, *My Childhood with Gandhi* (Ahmedabad: Navajivan Publishing House, 1957), 45; Dhupelia-Mesthrie, *Gandhi's Prisoner,* 74.

51. West, "Early Days."

52. Gandhi, *Childhood,* 18.

53. Lloyd I. Rudolph and Susanne Hoeber Rudolph, *Postmodern Gandhi and Other Essays: Gandhi in the World and at Home* (Chicago: University of Chicago Press, 2006).

54. James D. Hunt, "Experiments in Forming a Community of Service: The Evolution of Gandhi's First Ashrams, Phoenix and Tolstoy Farm," in *World Problems and Human Responsibility: Gandhian Perspectives,* ed. K. L. Seshagiri Rao and Henry O. Thompson (New York: Unification Theological Seminary, 1988), 178-203.

55. Gandhi, *Childhood,* 55.

56. Ibid., 114-115.

57. Rummonds, *Nineteenth-Century Printing Practices,* 598.

58. Ibid., 265.

59. The piece by Albert West ("In the Early Days with Gandhi") has no pagination. In the printed edition this quotation appears on page 14. The discussion of West's construction of "the wheel" is on page 6.

60. S. N. Bhattacharya, *Mahatma Gandhi: The Journalist* (Bombay: Asia Publishing House, 1965), 118.
61. Letterheads from NAD CSO 1758, 1904/2954; NAD CSO 1848, 8564/1907; NAD II 1/180 I 1058/1911.
62. Gandhi, *Childhood*, 46.
63. Millie Polak, *Mr. Gandhi: The Man* (London: George Allen and Unwin, 1931), 53-54.
64. Gandhi, *Autobiography*, 303-304.
65. Bhattacharya, *Mahatma Gandhi*, 113.
66. Gandhi, *Childhood*, 58.
67. Tony Ballantyne, *Orientalism and Race: Aryanism and the British Empire* (Basingstoke: Palgrave, 2002), 4; Antoinette Burton, *Brown over Black: Race and the Politics of Postcolonial Citation* (Delhi: Three Essays Collective, 2012); Jonathon Glassman, "Slower than a Massacre: The Multiple Sources of Racial Thought in Colonial Africa," *American Historical Review* 109, no. 3 (2004): 720-754; Jonathon Glassman, *War of Words, War of Stones: Racial Thought and Violence in Colonial Zanzibar* (Bloomington: Indiana University Press, 2011).
68. Soske, "'Wash Me Black Again,'" 194.

3. Indian Opinion

1. M. K. Gandhi, *An Autobiography or A Story of My Experiments with Truth*, trans. Mahadev Desai (Bombay: Navajivan, n.d.), 3730/3731 of 6567, e-book.
2. Quoted in S. N. Bhattacharya, *Mahatma Gandhi: The Journalist* (Bombay: Asia Publishing House, 1965), 79.
3. *Young India*, July 2, 1925, quoted in Bhattacharya, *Mahatma Gandhi*, 80.
4. This discussion is based on the book pages from the following editions of *Indian Opinion*: November 28, 1908; July 31, 1909; February 26, 1910; November 26, 1910; March 25, 1911; November

23, 1912; December 28, 1912; January 4, 1913; March 22, 1913; April 12, 1913; July 26, 1913; August 23, 1913; April 22, 1914. Each book pages contains both English and Gujarati, the latter having been professionally translated.

5. There is an extensive scholarship in this area. I have found the following useful: Ann Ardis and Patrick Collier, eds., *Transatlantic Print Culture, 1880–1940* (Basingstoke: Palgrave Macmillan, 2008); Margaret Beetham, "Open and Closed: The Periodical as a Publishing Genre," *Victorian Periodicals Review* 22, no. 3 (1989): 96–100; Faith Binckes, "Lines of Engagement: *Rhythm*, Reproduction, and the Textual Dialogues of Early Modernism," in *Little Magazines and Modernism: New Approaches*, ed. Suzanne W. Churchill and Adam McKible (Aldershot: Ashgate, 2007), 21–34; Laurel Brake, "Star Turn? Magazine, Part-Issue, and Book Serialisation," *Victorian Periodicals Review* 34, no. 3 (2001): 208–227; Laurel Brake and Julie F. Codell, eds., *Encounters in the Victorian Press: Editors, Authors, Readers* (Basingstoke: Palgrave Macmillan, 2010); Laurel Brake and Anne Humphreys, "Critical Theory and the Periodical," *Victorian Periodicals Review* 22, no. 3 (1989): 94–95; Suzanne W. Churchill and Adam McKible, eds., "Introduction," *Little Magazines and Modernism*, 1–18; Paul Keen, "Foolish Knowledge: The Commercial Modernity of the Periodical Press," *European Romantic Review* 19, no. 3 (2008): 199–218; Jon P. Klancher, *The Making of English Reading Audiences, 1790–1832* (Madison: University of Wisconsin Press, 1987); Mark Schoenfield, *British Periodicals and Romantic Identity: The "Literary Lower Empire"* (Basingstoke: Palgrave Macmillan, 2009).

6. Uma Dhupelia-Mesthrie, "The Place of India in South African History: Academic Scholarship, Past, Present and Future," *South African Historical Journal* 57 (2007): 12–34; Uma Dhupelia-Mesthrie, *Gandhi's Prisoner? The Life of Gandhi's Son Manilal* (Cape Town: Kwela, 2004); Uma Shashikant Mesthrie, "From Advocacy to Mobilization: *Indian Opinion*, 1903–1914," in *South Africa's Alterna-*

tive Press: Voices of Protest and Resistance, 1880s–1990s, ed. Les Switzer (Cambridge: Cambridge University Press, 1987); see also B. Pachai, "The History of the 'Indian Opinion' 1903-1914," in *Archives Yearbook for South African History* (Cape Town: Office of the Director of Archives, 1963), 1–127c.

7. David M. Henkin, *The Postal Age: The Emergence of Modern Communications in Nineteenth-Century America* (Chicago: University of Chicago Press, 2006).

8. M. K. Gandhi, *Satyagraha in South Africa* (Ahmedabad: Navajivan, 1928), 92.

9. Surendra Bhana and Neelima Shukla-Bhatt, eds., *A Fire that Blazed in the Ocean: Gandhi and the Poems of Satyagraha in South Africa, 1909–1911* (New Delhi: Promilla and Bibliophile South Asia, 2011), 37.

10. These included pen names (A. Chessel Piquet for Polak), initials (A. H. W. for Albert West), and correspondents, in some cases fictitious—"Our Own Correspondent" being a staff member collecting subscriptions in another town.

11. For reports by Polak, see *Indian Opinion* August 7 and 14, 1909; September 18 and 25, 1909; October 30, 1909. For West, see October 20, 1906; January 26, 1907.

12. For examples, see *Indian Opinion* November 19, 1903; August 20, 1903; March 28, 1908; July 3, 1909.

13. Reuter-sourced stories pick up from late 1906, presumably as a result of Polak taking over as editor.

14. "London Letter" and "Durban Notes" were regular features in the English section, the first starting on April 1, 1905, as "Our Weekly London Letter," then as "Our London Letter," until the 1920s. The second ran for two years, from late 1906 until October 1908. "Johannesburg Letter" was Gandhi's column in the Gujarati section that ran from 1906 until 1909. "Silhouettes" ran on August 4 and 24, 1907. *Indian Opinion* made frequent reference to the "Man in the Moon" columns: May 13, 1905; July 15, 1905; November 2, 1907; January 15, 1910; February 24, 1914.

15. On Sindhi merchants, see Claude Markovits, *The Global World of Indian Merchants, 1750–1947: Traders of Sind from Bukhara to Panama* (Cambridge: Cambridge University Press, 2000).

16. For reports on Gokhale's "protracted" arrival and departure, see *Indian Opinion* September 14, 1912; September 21, 1912; September 28, 1912; October 26, 1912; November 2, 1912; November 9, 1912; December 14, 1912; January 11, 1913; January 25, 1913. See also the pamphlet *Indian Opinion Souvenir of the Hon. Gopal Krishna Gokhale's Tour in South Africa, October 22nd to November 18th* (Durban: Indian Opinion, 1912).

17. For representative samples, see *Indian Opinion*: British Columbia, January 28, 1904, April 22, 1905, April 6, 1907, September 21, 1907, June 16, 1906, June 3, 1911; New Zealand, April 17, 1909, July 22, 1911.

18. As one reader of this manuscript pointed out, one trajectory for this reading community might have been the Gandhi-Polak household in Johannesburg in 1905–1906, which included readings of religious and ethical texts each night after dinner.

19. Henry David Thoreau, *Life without Principle*, 1854, http://en .wikisource.org/wiki/Life-Without_Principle (accessed September 15, 2011).

4. Binding Pamphlets, Summarizing India

1. M. K. Gandhi, *Satyagraha in South Africa* (Ahmedabad: Navajivan, 1928), 92.

2. Peter Stallybrass, "What Is a Book?" (lecture, Centre for the Study of the Book, Bodleian Library, University of Oxford, April 13, 2010).

3. Joseph Lelyveld, *Great Soul: Mahatma Gandhi and His Struggle with India* (New York: Knopf, 2011), 64.

4. John Kelly, "Fiji's Fifth Veda: Exile, Sanatan Dharm, and Counter-colonial Initiatives in Diaspora," in *Questioning Ramayanas: A South Asian Tradition,* ed. Paula Richman (New Delhi: Oxford University Press, 2000), 329–351.

5. Philip Lutgendorf, *Hanuman's Tale: The Messages of a Divine Monkey* (New York: Oxford University Press, 2007), 125. Despite Hanuman belonging to the Vaishnavite family, he found enthusiastic support among a largely Shaivite indentured community; see Alleyn Diesel, "Hinduism in KwaZuluNatal, South Africa," in *Culture and Economy in the Indian Diaspora,* ed. Bhikhu Parekh, Gurharpal Singh, and Steven Vertovec (London: Routledge, 2003), 42. For accounts of Katha and Jhunda ceremonies featuring Hanuman in Natal, see Sabita Jithoo, "Structure and Developmental Cycle of the Hindu Joint Family" (dissertation, University of Natal, 1970), 149–150; Ranj S. Nowbath, "The Hindus of South Africa," in *The Hindu Heritage in South Africa,* ed. Ranj S. Nowbath, Sookraj Chotai, and B. D. Lalla (Durban: The South African Hindu Maha Sabha, 1960), 17–22.

6. For information on Erasmus from his death records, see SAD MHG 5111/45. On Boer prisoners of war, see Elria Wessels, *Bannelinge in die Vreemde* (Pretoria: Kraal-uitgewers, 2010).

7. Rajend Mesthrie, *Language in Indenture: A Sociolinguistic History of Bhojpuri-Hindi in South Africa* (Johannesburg: Witwatersrand University Press, 1991).

8. Gandhi, *Satyagraha,* 91.

9. For colonial surveillance of *Indian Opinion,* see NAD II I/128, II I/135, II I/138; NAD IRD 45 1215/1905; SAD PM 42 1903/1767.

10. J. L. P. Erasmus, *The Story of the Ramayana: The Epic of Rama Prince of India* (Phoenix: International Printing Press, 1906).

11. Pyarelal, *The Birth of Satyagraha: From Petitioning to Passive Resistance,* vol. 3 (Ahmedabad: Navajivan, 1986), 87.

12. See also John D. Kelly and Martha Kaplan, "Diaspora and Swaraj, Swaraj and Diaspora," in *From the Colonial to the Postcolonial: India and Pakistan in Transition,* ed. Dipesh Chakrabarty, Rochona Majumdar, and Andrew Sartori (New Delhi: Permanent Black, 2007), 311–331; Uma Mesthrie, "Reducing the Indian Population to a 'Manageable Compass': A Study of the South African Assisted Emigration Scheme of 1927," *Natalia* 15 (1985): 36–56.

13. H. S. L. Polak, "The Transvaal Indians," *Modern Review* 7, no. 5 (1910): 422–426.

14. Uma Shashikant Mesthrie, "From Advocacy to Mobilization: *Indian Opinion*, 1903–1914," in *South Africa's Alternative Press: Voices of Protest and Resistance, 1880s–1990s*, ed. Les Switzer (Cambridge: Cambridge University Press, 1987), 120.

15. See Gandhi's comments on Tulsidas's translation of Valmiki: "Really speaking, Tulsidasa's work is not a translation. His devotion to God was so profound that instead of translating, he poured forth his own heart" (*CW* 9: 202). Another comment on the same theme pertains to a mass meeting where Hajee Habib spoke: "His speech was so caustic and impassioned that even those who did not know Gujarati said they could follow its purport" (*CW* 5: 359).

16. H. S. L. Polak, *A Book and Its Misnomer* (Phoenix: International Printing Press, 1907).

17. M. K. Gandhi, *Hind Swaraj or Indian Home Rule,* ed. Anthony J. Parel (Cambridge: Cambridge University Press, 2009).

18. L. E. Neame, *The Asiatic Danger in the Colonies* (London: George Routledge & Sons, 1907).

19. On Neame, see F. R. Metrowich, "Lawrence Elwin Neame," in *Dictionary of South African Biography,* vol. 4, ed. C. J. Byers (Pretoria: Human Sciences Research Council, 1968), 400. Neame attended the welcoming banquet for Gokhale in Johannesburg (*Indian Opinion,* November 9, 1912).

20. Neame, *Asiatic Danger,* viii.

21. Ibid., 9.

22. Polak, *A Book and Its Misnomer,* 3–5.

23. Ibid., 11.

24. Ibid., foreword, 2.

25. Ibid., 12.

26. These pamphlets comprise a biography and a speech. The former appears in the *Collected Works,* but the latter does not, and it also appears to have been lost in its pamphlet form.

27. Uday S. Mehta, "Patience, Inwardness, and Self-Knowledge in Gandhi's *Hind Swaraj*," *Public Culture* 23, no. 2 (2011): 417-429.

28. John Ruskin, *Unto This Last: Four Essays on the First Principles of Political Economy,* 1862 (London: George Allen, 1893), 65.

29. Ibid., 22.

30. Ibid., 23.

31. Catherine Gallagher, *The Body Economic: Life, Death, and Sensation in Political Economy and the Victorian Novel* (Princeton, N.J.: Princeton University Press, 2006), 88.

32. Ibid., 90.

33. Ruskin, *Unto This Last,* 65.

34. Gandhi, *Hind Swaraj.*

5. A Gandhian Theory of Reading

1. Anthony J. Parel, "Editor's Introduction," in *Hind Swaraj and Other Writings,* ed. Anthony J. Parel (Cambridge: Cambridge University Press, 2009), xiii–lxii.

2. Ibid., xiv.

3. M. K. Gandhi, "Preface to the English Translation," in *Indian Home Rule or Hind Swaraj. In Hind Swaraj and Other Writings,* 5–8.

4. Anthony Parel, "The Origins of *Hind Swaraj,*" in *Gandhi and South Africa: Principles and Politics,* ed. Judith M. Brown and Martin Prozesky (Pietermaritzburg: University of Natal Press, 1996), 35–66, provides an excellent discussion of the South African context of *Hind Swaraj.* His interests focus on the philosophy of Gandhi and so do not touch on the figure of the reader of *Indian Opinion.*

5. Keith Breckenridge, "Gandhi's Progressive Disillusionment: Thumbs, Fingers, and the Rejection of Scientific Modernism in *Hind Swaraj,*" *Public Culture* 23, no. 2 (2011): 331–348; Joseph Lelyveld, *Great Soul: Mahatma Gandhi and His Struggle with India* (New York: Knopf, 2011); Claude Markovits, *The Un-Gandhian Gandhi: The Life and Afterlife of the Mahatma* (New Delhi: Permanent Black, 2003);

Maureen Swan, *Gandhi: The South African Experience* (Johannesburg: Ravan Press, 1985).

6. M. K. Gandhi, *Satyagraha in South Africa* (Ahmedabad: Navajivan, 1928), 92.

7. Quoted in S. N. Bhattacharya, *Mahatma Gandhi: The Journalist* (Bombay: Asia Publishing House, 1965), 107.

8. Surendra Bhana and Neelima Shukla-Bhatt, eds., *A Fire that Blazed in the Ocean: Gandhi and the Poems of Satyagraha in South Africa, 1909–1911* (New Delhi: Promilla and Bibliophile South Asia, 2011); Devarakshanam Betty Govinden, "The Mahatma, the Text and the Critic—in South Africa," *Scrutiny2* 13, no. 2 (2008): 47–62.

9. Bhana and Shukla-Bhatt, *Fire that Blazed.*

10. Ibid., 18.

11. Ibid., 178.

12. Ibid., 155–156.

13. Uday S. Mehta, "Patience, Inwardness, and Self-Knowledge in Gandhi's *Hind Swaraj*," *Public Culture* 23, no. 2 (2011): 417–429.

14. These views may also have drawn on Henry David Thoreau's ideas on reading in *Walden,* Kindle edition, www.gutenberg.org/ebooks /205 (accessed November 15, 2011). He contrasts those with "unwearied gizzard" who "vegetate and dissipate their faculties in what is called easy reading" with serious reading to which we must "devote our most alert and wakeful hours" (1345/4671; 1331/4671).

15. John Ruskin, *Sesame and Lilies,* 1865 (n.p.: Dodo Publishers, n.d.), 1.

16. Ibid., 8–9.

17. Ibid., 9.

18. Ibid., 19.

19. Ibid., 35.

20. Ananda M. Pandiri, *A Comprehensive, Annotated Bibliography on Mahatma Gandhi,* vol. 1 (Westport, Conn.: Greenwood Press, 1995), 298.

21. Ruskin, *Sesame,* 10.

22. M. K. Gandhi, "Foreword," in *Hind Swaraj and Other Writings,* ed. Anthony J. Parel (Cambridge: Cambridge University Press, 2009), 9-11.

23. Ibid., 10.

24. Ibid.

25. Ibid.

26. Ibid., 11.

27. Gandhi, "Foreword," in *Hind Swaraj,* 13.

28. Ibid.

29. Ibid., 101.

30. Ibid., 35-36.

31. Ibid., 36.

32. Ibid., 111.

33. Details for these *Indian Opinion* articles are as follows: "What Is Habitation?" and "Are Indian Peoples British Subjects?" (February 4, 1904); "What Is Hawking?" (September 17, 1904); "Are British Subjects Serfs?" (August 19, 1905); "What Is a Hawker?" and "Is a Verandah a Shop?" (October 14, 1905); "What Is a Coloured Person?" (August 8, 1908). Further examples include "What Is Colour?" (August 8, 1908); "What Is a Wife?" (April 20, 1912); "What Is an Asiatic?" (May 17, 1913); "When Is an Asiatic a European?" (May 31, 1913); "What Is Domicile?" (August 23, 1913); "What Is Patriotism?" (September 23, 1914).

34. *CW* 7: 330; *CW* 8: 41-44, 57-59, 94-96, 428-430, 438-442; *CW* 9: 13-16, 50-52, 206-207, 216-217; *CW* 6: 436-437; *CW* 5: 356-362.

35. Surendra Bhana and Kusum K. Bhoola, *Introducing South Africa or Dialogue of Two Friends, by an Indian* (Durban: Local History Museum, 2005).

36. Christopher Gill, "Afterword: Dialectic and the Dialogue Form in Late Plato," in *Form and Argument in Late Plato,* ed. Christopher Gill and Mary Margaret McCabe (Oxford: Clarendon Press, 1996), 283-311.

37. Gandhi, "Preface," in *Indian Home Rule,* 6.

38. M. K. Gandhi, *Hind Swaraj or Indian Home Rule*, ed. Anthony J. Parel (New Delhi: Cambridge University Press, 2009), 14.

39. Ibid., 15–16.

40. Ibid., 17, 19.

41. Ibid., 26.

42. Ibid., 80.

43. Ibid., 35–36.

44. Ibid., 112.

45. Ibid., 74.

46. Ibid., 32.

47. Mehta, "Patience."

48. Ibid., 422.

49. Plato, *Phaedrus*, trans. Robin Waterfield (Oxford: Oxford University Press, 2002).

50. Ibid., 70.

51. Christopher Gill, "The Platonic Dialogue," in *A Companion to Ancient Philosophy*, ed. Mary Louise Gill and Pierre Pellegrin (Oxford: Blackwell, 2008), 136–150.

Conclusion

1. S. N. Bhattacharya, *Mahatma Gandhi: The Journalist* (Bombay: Asia Publishing House, 1965), 37, 59.

2. Ibid., 56.

3. Ibid., 59.

4. Ananda M. Pandiri, *A Comprehensive, Annotated Bibliography on Mahatma Gandhi*, vol. 1 (Westport, Conn.: Greenwood Press, 1995), 155–242.

5. Anthony Parel's exemplary edition reinstates the text of the first English edition but omits the phrase "No Rights Reserved" from the title page; see *Hind Swaraj and Other Writings*, ed. Anthony J. Parel (Cambridge: Cambridge University Press, 2009). However, since the first edition of Parel's book appeared in 1997, when

Gandhi's work was still under copyright, this omission is understandable. As indicated earlier, Parel's focus is on unraveling the strands of Gandhi's philosophical thought, and so his interest does not encompass the reader of *Indian Opinion*. A third element also appears on the original title page of the first English edition and not in Parel's edition. This consists of the publishing details that read "The International Printing Press, Phoenix, Natal, 1910." However, since Parel's is not a facsimile edition, this omission is to be expected, and the publishing history of the text is masterfully covered in "A Note on the History of the Text" (lxiii).

6. I am grateful to Achal Prabhala for this point ("Would Gandhi Have Been a Wikipedian?" *Indian Express,* January 17, 2012, www.indianexpress.com/news/would-gandhi-have-been-a-wikipedian/900506/ [accessed January 28, 2012]).

7. Isabel Hofmeyr, "Violent Texts, Vulnerable Readers: *Hind Swaraj* and Its South African Audiences," *Public Culture* 23, no. 2 (2011): 285–297.

8. David Shields, *Reality Hunger: A Manifesto* (London: Penguin, 2010).

9. Luc Sante, "The Fiction of Memory," *Sunday Book Review,* March 12, 2010, www.nytimes.com/2010/03/14/books/review/Sante-t.html?pagewanted=all (accessed January 3, 2012).

10. Ellen Gruber Garvey, "Scissorizing and Scrapbooks: Nineteenth-Century Reading, Remaking, and Recirculating," in *New Media 1740–1915,* ed. Lisa Gitelman and Geoffrey B. Pingree (Cambridge, Mass.: MIT Press, 2003), 207–227.

11. Jean Comaroff and John L. Comaroff, *Theory from the South: Or, How Euro-America Is Evolving Towards Africa* (New York: Paradigm Publishers, 2011); see also Tamar Garb, in conversation with Achille Mbembe, Sarah Nuttall, Raison Naidoo, and Colin Richards, "Thinking from the South: Reflections on Image and Place," in *Figures and Fictions: Contemporary South African Photography,* ed. Tamar Garb (London: Steidl, 2011), 300–307.

12. Prabhala, "Wikipedian."

13. "Gandhi-Home-Rule-First-Edition-1909," en.wikipedia.org/wiki/
    File:Gandhi-Home-Rule-First-Edition-1909.jpg (accessed
    January 23, 2012).

14. Paul Cilliers, "On the Importance of a Certain Slowness," *E:CO* 8,
    no. 3 (2006): 105–112.

15. Henry David Thoreau, *Life without Principle* (1854), http://en
    .wikisource.org/wiki/Life-Without_Principle (accessed September
    15, 2011).

16. Leah Price, *The Anthology and the Rise of the Novel: From Richardson to
    George Eliot* (Cambridge: Cambridge University Press, 2000),
    e-book, 2167/3213, 89/3213, 2175/3213.

17. For exceptions, see Devarakshanam Betty Govinden, "The
    Mahatma, the Text and the Critic—in South Africa," *Scrutiny2* 13,
    no. 2 (2008): 47–62; Sunil Khilnani, "Gandhi and Nehru: The Uses
    of English," in *A History of Indian Literature in English*, ed. Arvind
    Krishna Mehrotra (London: Hurst and Company, 2003), 135–156;
    Madhumita Lahiri, "Gandhian Fictions: Rereading *Satyagraha in
    South Africa*" (paper presented to the School of Literature and
    Language Seminar, University of the Witwatersrand, June 2, 2011).

18. Bhabani Bhattacharya, *Gandhi the Writer* (New Delhi: National
    Book Trust, 1969), 1.

19. Priyamvada Gopal, *The Indian English Novel: Nation, History, and
    Narration* (Oxford: Oxford University Press, 2009), 43–68; Amaresh
    Datta, *The Encyclopedia of Indian Literature*, vol. 2 (Delhi: South Asia
    Books, 1988), 1350–1351.

A Note on Sources

In writing this book, my primary sources have been the English sections of *Indian Opinion* and the back translations of the Gujarati sections written by Gandhi for the newspapers and available in the *Collected Works (CW)* (both of these for the years 1903 to 1914, when Gandhi left for India). I do not read Gujarati, and hence I am working with translated texts, and with all the difficulties that these imply. There are further problems as well. Gandhi's writings made up probably about 20 percent of the Gujarati section, the length of which varied across the years from four to sixteen pages. Self-evidently the surrounding columns will have recontextualized and possibly contradicted Gandhi's columns.

These difficulties notwithstanding, the project remains worth pursuing. First, scholars rely more on the *CW* than on *Indian Opinion,* an imbalance that this study seeks to shift. Second, many analyses that use *Indian Opinion* treat writing from the English section and the translations of Gandhi's Gujarati writings in the *CW* as one field. These discussions assume further that *Indian Opinion* is a newspaper in a way that we understand that term today. This book charts a different course, starting with the premise that

Indian Opinion is a multilingual publication that combined features of the newspaper and the periodical in unique ways (and, throughout, these terms are used interchangeably). Within this framework, I treat the English section and Gujarati columns (via the translations) as related but discrete domains and attempt to bring to bear a literary perspective on these writings in contrast to the canonical Gandhian scholarship that uses these simply as social science "sources." The topic of literary approaches to the Gandhian corpus is addressed in the Conclusion.

In addition to the references in the endnotes, the following publications were consulted:

Periodicals

Hindi, 1921–1925

Indian Emigrant, 1914–1927 (incomplete run)

Indian Opinion, 1903–1914

Indian Printers Journal, 1912

Indian Social Reformer, 1894–1897

Indian Review, 1901–1925

Indian Views, 1914–1921

Modern Review, 1908–1928

South African Printer and Stationer, 1923–1927

Papers

Benarsidas Chaturvedi Papers, 321, National Archives of India, Delhi.

Chaturvedi Collection, Premchand Archives and Literary Centre, Jamia Millia Islamia University, Delhi.

Papers of the National Industrial Council of the Printing and Newspaper Industry of South Africa, KAD, A2509, Cape Town, Western Cape Provincial and Records Service.

ACKNOWLEDGMENTS

In keeping with its themes, this book has been made slowly. Its dim beginnings stretch back to 2002, when I first visited India. Like most traveling scholars, my trip was facilitated by the ever-generous Neeladri Bhattacharya. In India an intellectually hospitable community of scholars assisted me. Special thanks go to Rimli Bhattacharya, Nonica Datta, Chitra Joshi, Uday Kumar, Simi Malhotra, Tanika Sarkar, and Sumit Sarkar. Throughout, P. K. Datta has been a friend without peer. From Lakshmi Subramanian I have learned much. Achal Prabhala, a superb transnational comrade, helped this book along in various ways, including through an introduction to Sharmila Sen, whose instantaneous belief in this project has been inspiring. During a three-month stay in Bangalore I was hosted at the Centre for the Study of Culture and Society in 2005: thanks to Teju Niranjana, Rochelle Pinto, Ashish Rajadhyaksha, and Mrinalini Sebastian.

In more than a decade of working on Indian Ocean matters, I have encountered a marvelous republic of scholars: Sunil Amrith, Gabeba Baderoon, Sharad Chari, Yvette Christiansë, Bodil Folke Frederiksen, Mark Ravinder Frost, Devleena Ghosh, Stephanie Jones, Preben Kaarsholm, Michael Pearson, Meg Samuelson, Raymond

Suttner, and Nigel Worden. Antoinette Burton, Tony Ballantyne, and Tanya Agathocleous share Indian Ocean interests and generously commented on parts of the manuscript.

Betty Devarakshanam Govinden has been a colleague of long standing from whom I have learned much. Goolam Vahed, another wonderful scholar, shared his voluminous knowledge with me on many occasions. I owe a special word of thanks to Uma Dhupelia-Mesthrie, the preeminent figure in the study of South African Indian history: she has extended to me on many occasions her generosity, forbearance, extraordinary expertise, and archival material.

The Centre for Indian Studies in Africa and its earlier incarnations provided a congenial forum to try out some of the ideas in this book: my thanks to Denise Newfield, Stephen Gelb, Pamila Gupta, Madumitha Lahiri, Yvonne Reid, Jon Soske, Rehana Vally-Moosa, Michelle Williams, and the late and continually mourned Pippa Stein. Dilip Menon, who made the Indian Ocean crossing from Delhi to Johannesburg, has been an inspirational colleague. Yunus Ballim, Belinda Bozzoli, Tawana Kupe, and Loyiso Nongxa vouchsafed university support for the various South Africa–India ventures that fed both directly and indirectly into this book. It was a pleasure working with Navdeep Suri and Vikram Doraiswamy, both outstanding consuls general of India in Johannesburg.

A collegial community of scholars at the University of the Witwatersrand makes intellectual life interesting. Thanks to Michelle Adler, Prinisha Badassy, Phil Bonner, Keith Breckenridge, Cathy Burns, Jackie Cock, Lesley Cowling, Liz Gunner, Shireen Hassim, Judith Inggs, Mark Leon, Achille Mbembe, Libby Meintjes, Ashlee Neser, Vish Satgar, and Sue van Zyl. Merle Govind, Pumla Dineo Gqola, Litheko Modisane, James Ogude, Dan Ojwang, and Bheki

ACKNOWLEDGMENTS

Peterson have been patient and generous colleagues. Sarah Nuttall's friendship and conversations have enriched this book.

In London Annie Coombes, Deborah James, Wayne Dooling, and Ruth Watson have afforded friendship and hospitality for the weary traveler. In Cape Town Carolyn Hamilton and Pippa Skotnes have been patient friends and colleagues.

Heather Hughes kindly provided information on John Dube, and Mark Sandham showed me how hand printing worked. Sydney Shep generously made available to me digital copies of the *Indian Printers' Journal* and read parts of the manuscript. Surendra Bhana and Neelima Shukla-Bhatt kindly assisted with translating and transliterating Gujarati titles.

Frances Williams's professional wisdom has helped me and the making of this book enormously.

This book has benefited from the support of friends: Ruth Becker, Helen Becker and Michael Harmey, Bridget Lamont, Ros Lamont, Moggie Lewis, Karen Lazar, David Medalie, Sandy Prosalendis, John and Alice Parkington, Catherine Stewart, and Estelle Trengove. Thanks to my writing companions, Karen Martin, Catherine Garson, and Annie Holmes, and to Helen Struthers and ANOVA for the generous use of an office.

Elise and Kevin Tait and Jan and Angela Hofmeyr and their respective "tribes" have buoyed me up and along, as has Mary Park.

Parts of this book were presented at seminars in Bangalore, Boston, Cambridge, Cape Town, Delhi, Durban, Grahamstown, Johannesburg, Kampala, Kolkata, London, Madison, Michigan, Mumbai, New York, Oxford, Pretoria, and Zanzibar. My thanks to the hosts, organizers, and audiences of these various events.

The National Research Foundation, the University of the Witwatersrand, and the Danish Council for Independent Research–Humanities funded aspects of the research for this book. A month's fellowship at the Stellenbosch Institute for Advanced Studies, Wallenberg Research Centre at Stellenbosch University, allowed me to make significant headway with this project.

A small portion of Chapter 1 and portions of Chapter 2 are reprinted by permission from "Gandhi's Printing Press: Indian Ocean Print Cultures and Cosmopolitanisms," in *Cosmopolitan Thought Zones: South Asia and the Global Circulation of Ideas*, ed. Sugata Bose and Kris Manjapra (London: Palgrave Macmillan, 2010), 112–130. Brief portions of Chapter 5 are reprinted by permission from "Violent Texts, Vulnerable Readers: *Hind Swaraj* and Its South African Audiences," *Public Culture* 23, no. 2 (2011): 285–297.

This book is dedicated to the memory of my dear friend, Miriam Abrams, who taught us all how to live in this world and how to depart it with grace and courage.

Index

CPSIA information can be obtained
at www.ICGtesting.com
Printed in the USA
LVHW092340260821
696229LV00002B/23/J